POLITICAL RESPONSIBILITY & TECH GOVERNANCE

Not a day goes by without a new story on the perils of technology: from increasingly clever machines that surpass human capability, comprehension and control to genetic technologies capable of altering the human genome in ways we cannot predict. How can we respond? What should we do politically? Focusing on the rise of robotics and artificial intelligence (AI), and the impact of new reproductive and genetic technologies (Repro-tech), Jude Browne questions who has political responsibility for the structural impacts of these technologies and considers how we might go about preparing for the far-reaching societal changes they may bring.

Jude Browne is Head of the Department of Politics and International Studies at the University of Cambridge, a Professorial Fellow in Social and Political Sciences at King's College and the Frankopan Director of the University of Cambridge Centre for Gender Studies. Browne's research interests are in political and feminist theories of equality, political responsibility, public interest, public policy, structural injustice and the impact of technology on society.

POLITICAL RESPONSIBILITY & TECH GOVERNANCE

AI, Repro-tech and Structural Injustice

Jude Browne

University of Cambridge

CAMBRIDGE
UNIVERSITY PRESS

Shaftesbury Road, Cambridge CB2 8EA, United Kingdom

One Liberty Plaza, 20th Floor, New York, NY 10006, USA

477 Williamstown Road, Port Melbourne, VIC 3207, Australia

314–321, 3rd Floor, Plot 3, Splendor Forum, Jasola District Centre,
New Delhi – 110025, India

103 Penang Road, #05-06/07, Visioncrest Commercial, Singapore 238467

Cambridge University Press is part of Cambridge University Press & Assessment,
a department of the University of Cambridge.

We share the University's mission to contribute to society through the pursuit of
education, learning and research at the highest international levels of excellence.

www.cambridge.org
Information on this title: www.cambridge.org/9781009447355

DOI: 10.1017/9781009447362

When citing this work, please include a reference to the DOI 10.1017/9781009447362

First published 2025

A catalogue record for this publication is available from the British Library

A Cataloging-in-Publication data record for this book is available from the Library of Congress

ISBN 978-1-009-44735-5 Hardback

For Umar, Etta & Martha
And for my mother, Elizabeth 1935–2024

Technology is the answer, but what was the question?
(Cedric Price, 1966)

Contents

Preface

How should we think politically about the ways in which transformative technologies such as artificial intelligence (AI) and reproductive and genetic technologies (Repro-tech) structurally change society? Considering this question through the concept of structural injustice, I explore how we might begin to prepare for the potentially seismic societal changes these technologies promise to bring.

Acknowledgements

I want to thank so many people for encouraging me to write this book and for helping me craft the ideas that populate it over the past five years. For a sizeable chunk of this time I have been Head of the Department of Politics and International Studies at the University of Cambridge – a phenomenally supportive, collegiate and intellectually dynamic department, and I am indebted to my colleagues, both academic and professional, for all their support and encouragement along the way, as well as to the many brilliant students who engaged with my ideas in my lectures, seminars and supervisions – for me, this is what being an academic is all about.

There were also numerous audiences and interlocutors in both academic and public settings that provided immensely valuable constructive critical engagement. To name a few of these which were particularly key: 'Structural Injustice and the Regulatory Public Body Landscape', *Structural Injustice and the Law* (hosted by Virginia Mantouvalou (UCL) and Jonathan Wolff (Oxford). UCL, London, 28–29 March 2022); 'AI and Structural Injustice', Gender & Tech International Conference for Oxford University Press (21 September 2021 – conducted live on Zoom due to COVID). Proceedings published by Oxford University Press, 2024; Jude Browne in conversation with Amol Rajan, Hay Festival (27 June 2021); 'Putting the Public into Public Bodies', Penn State University School of Public Policy (discussant; Tony Bertelli, 21 June 2021); 'Structural Injustice, Traceability and Political Responsibility', International Workshop on Structural Injustice, University of Cambridge (co-hosted with Maeve McKeown and with Brooke Ackerly, Ryoa Chung, Monique Deveaux, Agomoni Ganguli-Mitra, Lewis Gordon, Sally Haslanger, Clarissa Hayward, Adam Hosein,

Alison Jaggar, Catherine Lu, Virginia Mantouvalou, Mara Marin, Alasia Nuti, Serena Parekh, Jade Schiff, Theresa Tobin and Timothy Waligore, 22 March 2021); Lay-governance – the public body. Lay-Governance Workshop University of Cambridge and Sciences Po CAMPO (co-hosted with Annabelle Laver and with Jane Mansbridge, Tony Bertelli, David Owen, Phil Parvin, Joseph Heath and John Boswell, 12 April 2021); 'Political Responsibility', Contemporary Political Theory Seminar, University of Cambridge (chaired by Duncan Bell, Cambridge, 31 January 2020); 'Technology, Gender & the Reproductive Habits of Future Generations', University of Cambridge London Engagement Series 2019, 'Structural Injustice and the Public Interest', Ethics and Public Policy Conference, Blavatnik School of Government, University of Oxford (chaired by Jo Wolff, 7 June 2019); 'Gender & Tech', University of Cambridge (chaired by Stephen Toope, Hong Kong, 30 April 2019); 'Political Responsibility and the Public Interest', Department of Political Science, Yale University (supported by the European Studies Fund, chaired by Professor Seyla Benhabib, The Eugene Meyer Professor of Political Science and Philosophy, 14 November 2018); '100 Years to Bliss? AI, Politics and Regulation', The Centre for the Future of Intelligence, University of Cambridge (chaired by David Runciman, Cambridge, 27 September 2018). As I have been writing this book, parts of it or some of the ideas have been published in the following: Browne, J. (forthcoming 2024) Structural injustice and the regulatory public body landscape. In Virginia Mantouvalou and Jonathan Wolff (eds.) *Structural Injustice and the Law.* London: UCL Press; Browne, J. (2023) AI & structural injustice: A feminist perspective. In Jude Browne, Steven Cave, Eleanor Drage and Kerry McInerney (eds.) *Feminist AI: Critical Perspectives on Algorithms, Data, and Intelligent Machines.* Oxford: Oxford University Press, pp. 328–346: Browne, J. (2023) Political implications of the 'untraceability' of structural injustice. *Contemporary Political Theory.* Early View: https://link .springer.com/article/10.1057/s41296-023-00634-4. Browne, J. (2020) The regulatory gift: Politics, regulation and governance. *Regulation and Governance* 14(2): 165–388.

I owe a debt of gratitude to Helen Yanocuplous for so much encouragement and humorous commentary on the writing of this book.

Particular thanks also to Jason Sharman for reading early draft chapters and for his constant and invaluable encouragement with this project over the years; to Andrea Sangiovanni, Jonathan Wolff, Maeve McKeown, Duncan Bell, Sylvie Delacroix and Sheila Benhabib for the many, many hours of generous discussion on the theoretical arguments I make on structural injustices; to the members of my writing workshop group, Mette Eilistrup-Sangiovanni, Sarah Fine, Sylvie Delacroix and the others who joined us along the way; and to Ayse Zarakol, Devon Curtis and Melissa Calarescu for collectively providing me with the most productive writing periods in amongst the intensely busy terms of the academic year at Cambridge as well as for helping me think through the narrative of the book as it evolved over time. Special thanks to John Dunn for reading and discussing with me various drafts with excellent feedback and, most importantly, reassurance. I am extremely grateful to him for taking the time to discuss my ideas and to explore how they fit into current political debates.

Huge thanks to Emma King and Chloe Brown for helping me get this book over the line through their encouragement and support and to Llinos Edwards, an expert professional copy editor, and Carl Pierer, a PhD student in political theory at the University of Cambridge, who acted as my research assistant over the past two years. I had been thinking about the ideas in this book for so long that I often made assumptions in my writing that the reader could simply make the leaps from argument to argument without any explanatory bridge. Each of them, in their different ways, guided me to at least cast down a few stepping stones, and I owe them both a great debt of gratitude. Thanks also to Carrie Parkinson at Cambridge University Press, and to John Haslam, Executive Publisher, whose generous guidance, encouragement and patience were essential to the completion of this book. His great enthusiasm for bringing ideas about politics out into the world is infectious.

My deepest gratitude is to Umar, not only for his love, care and humour but for tolerating my obsessive monologues on why this idea or that idea worked or did not work and for helping me to wrestle with the most difficult parts. He deserves some sort of medal. Thanks too, to my extraordinary mother, Elizabeth, who was always a constant and wholly supportive sage; and finally, to my wonderful children, Etta and Martha, who spurred me on to try to find some answers . . ..

Introduction

NOT A DAY GOES BY WITHOUT A NEW STORY ON THE PERILS of technology. We hear of increasingly intelligent machines that surpass human capability, comprehension, and control. We hear of tech billionaires imploring each other to abandon artificial intelligence (AI) that poses 'profound risks to society'.[1] We hear of genetic technologies capable of altering the human genome in ways we cannot predict and which are leading us to a future two-tier humanity consisting of those of us who are genetically enhanced and those who are not. How can we respond to these stories? What should we do politically?

When the well-known futurist and technology expert, Amy Webb, was recently asked by interviewer Wajahat Ali what we should do to address tech-generated harm to human society, her answer, although surely right, nevertheless seemed impossibly impracticable:[2]

> WEBB: The challenge has always been *complacency*. The antidote to that is paying better attention. If you see something, say something – just like you are on the New York subway.

---

[1] For example, Max Tegmark, Professor of Artificial Intelligence & Fundamental Interactions at MIT, wrote an open letter entitled 'Pause Giant AI Experiments' which was signed by over 33,000 people including Elon Musk CEO of SpaceX, Tesla & X/ Twitter, Stuart Russell, Berkeley, Professor of Computer Science, Steve Wozniak, Co-founder, Apple to name just a few. See full letter here: https://futureoflife.org/open-letter/pause-giant-ai-experiments

[2] The interview was conducted by 'South by Southwest' host Wajahat Ali in 2023, and is paraphrased here for clarity. The full interview is available at: www.youtube.com/watch?v=54ZPJ3WDpPk&t=12s.

ALI: I'm not complacent – I'm exhausted, I'm overwhelmed! There's ... 'income inequality', there's 'climate change' ... I do not really understand this stuff. Look, I'm not a brilliant tech expert like you, I'm just an average Jane, Joe or José. Even if 'I see something and I say something', no one's going to care. What hope does the average person have of changing something?

WEBB: There is no one big solution – no big switch that can be flipped – but we can make a thousand incremental changes through the decisions that we are making every day. I know that everyone is exhausted, but it's what we have to do!

What Amy Webb is describing here is a huge political challenge. Rather than being able to point to anyone in particular as the root cause, this is a problem of vast arrays of complex, multitudinous 'everyday' actions of individuals, groups and institutions – otherwise known as structural dynamics. So, who should take up the task of deciding which micro-incremental changes to make, and coordinate them to sufficient effect? What tools can we develop to politically address structural challenges of this sort, especially when states are so often captured by the interests of tech industries and seduced by the promise of technological efficiency and betterment?

I argue that a radical rethink about how the substantial coordinating capabilities of state-level governance is devised and deployed is urgently required.

At the core of this book is an anxiety about the failure of democratic politics to govern transformative technologies.[3] By way of example, I have chosen to focus on AI and reproductive and genetic technologies

---

[3] People interested in writing about democratic politics in the neoliberal context are often motivated by a particular anxiety. For Wendy Brown (2019) it is the 'economization' of contemporary political thought; for Sara Ahmed (2021), it is the resilience of baleful institutional practices; for Danielle Allen (2023) it is the failure to create an inclusive politics for diverse, disconnected communities; for Anne Phillips (2021) it is the state's deviation from equality as the principal concept on which to ground politics; for David Runciman (2018) it is the political complacency surrounding the demise of democracy; for John Dunn (1990) it is a lack of prudence within political communities. These are of course closely connected, but in this book, I undertake to articulate why a focus on the relationship between structural harm and transformative technologies is important.

(Repro-tech) as two technologies set to dominate future societies. While there is a great deal of hype around these extraordinary scientific developments, the question remains of what sort of practical political methods could be deployed to address the harms they may bring alongside the great many advantages they promise. By way of exploring this question, I want to move beyond the usual arguments and legal devices that serve to identify tech developers, and users, as being at fault for individual acts of wrongdoing, recklessness, incompetence or negligence (important as these questions of liability are) and ask instead how we might address the broader structural dynamics intertwined with the increasing use of AI and Repro-tech. My argument will be that to take a much sharper structural perspective on these transformative technologies is a vital requirement of contemporary politics.

BOOK STRUCTURE

CHAPTER 1 THE PROBLEM OF STRUCTURAL INJUSTICE: WHERE NOBODY IS LIABLE, WHO IS RESPONSIBLE? Before engaging with the specific questions of how AI and Repro-tech relate to structural injustice and what this might mean for tech governance, I begin with a theoretical discussion of what structural injustice is and how it operates. This is an unusual place to start a book on tech governance but I argue that it has significant value for thinking differently about the challenges of addressing tech-generated harms to society.[4]

Through the process of writing this book and also as co-editor of *What is Structural Injustice?* (Browne and McKeown, 2024),[5] I came to appreciate that there were a great many different interpretations of the concept of structural injustice. At the centre of these debates, however, is the late Iris Marion Young's extraordinary text *Responsibility for Justice* (2011). The array of interpretations and criticisms inspired by Young's last work are

---

[4] As I shall discuss further on, I take the concept of structural harm and structural injustice to mean the same thing.

[5] In this volume, my co-editor, Maeve McKeown, and I came to different conclusions on the question of the book's title, *What Is Structural Injustice?* This made for an infinitely more interesting and productive set of discussions than if we had taken our interpretations of structural injustice in the same direction. I discuss this in more detail further on.

likely to be, in part at least, because it was published posthumously and based on drafts and notes, with several gaps and inconsistencies inevitable in any draft. Nevertheless, a wealth of scholarship emanating from engagement with this work has taken us in many different intellectual directions, opening up new theoretical and practical opportunities to address injustice. As I shall set out, I develop my own particular interpretation of structural injustice, in part based on my exploration of Young's work and that of her critics.

I begin by focusing on an overlooked thread in Young's work – the *untraceabilty* of structural dynamics, which although not tied to technology by Young, is, I suggest, a valuable lens through which to consider technology's impact on society. Not only do I see untraceability as a constant theme in Young's account of structural injustice, but also, in fact, a defining feature. This argument, I suggest, brings with it profound implications for the sort of politics we might employ to address the structural injustices intertwined in the rise of transformative technologies and which, I argue, are not catered to in current liability-based governance measures such as global AI Ethics Frameworks, General Data Protection Regulation (GDPR), licensing laws or international bans on human cloning, for example (as important as these sorts of approaches are).

Put simply, 'untraceability' in this context means that the causes of structural injustice are too complex and convoluted to be meaningfully traced to an agent of fault. This is not an observation about the ineptitude of our political commitment but rather one of political impossibility in a given time and place. This is not to say that there will never be new social scientific and technical tools in the future that will give us new insights, but rather to emphasise the political inefficacy of attempting to trace blame where none can be meaningfully traced.

As I shall explain, because structural injustice is conceptually distinct from other sorts of injustice, to turn our political machinery, in whatever form, towards tracing culpable agents for the purpose of addressing such injustice (what Young called the 'liability model of responsibility'),[6] is

---

[6] Young (2011: 97). I personally find the term 'fault' or 'guilt' (direct or indirect) clearer than 'liability' but because I am drawing on Young's work, I use the term 'fault-based' or 'guilt-based' interchangeably with 'liability-based' injustice throughout the book.

misguided. We need a different kind of political sensibility to address structural injustices.

In order to set the ground for my arguments later in this book on the structural dynamics of AI and Repro-tech, I take time in Chapter 1 to pull out the elements of Young's work that offer a persuasive account of the untraceability of structural injustice and I consider the ways in which, influenced by her critical engagement with the work of Hannah Arendt, Young leads us to the question of '*where nobody is liable, who is responsible?*' This question, I argue, is important in the context of tech governance and as such, serves as a central theme of the book

Many intellectual disputes with Young's work are built on the seemingly logical question that if we cannot conceivably trace responsibility for structural injustice then how can we claim that any particular individual, group or institution is politically responsible for addressing structural injustice? This is certainly an important question, and in the course of the book I shall try to explain why I think we need to look beyond liability in developing a more progressive politics around the structural dynamics of transformative technologies.

I should stress, however, that while the book is in places a dialogue with the work of Young, I do not view this as an exercise in discovering Young's ultimate meaning. The primary intention is not to analyse, critique or defend Young's work per se. I freely admit that my version of structural injustice is built on a purposefully selective reading of Young's arguments, and consequently I create a much sharper version of structural injustice than may be found in Young's work or in the work of other thinkers on structural injustice. Nevertheless, Young's claim that remedies for structural injustice ought not to rely on 'traceable liability' is the direction I develop and advocate in this book because I see it as a way to address forms of injustice and harm that are often missed by a politics too heavily weighted towards liability.[7]

---

[7] For those readers who are less interested in the political theory arguments behind the latter arguments in this book on governance of transformative technologies in the context of structural injustice, I have provided a summary of concepts in Appendix 1 and it is possible to read this book with a sharper focus on technology and governance from Chapter 2 onwards.

Young's approach was not to construct an entire new theory of political responsibility for the unintended structural consequences of human behaviour, but rather to adopt an approach of 'pragmatic theorizing' (1997: 17). The theoretical activity of such an approach 'is not concerned to give an account of the whole' but rather is constituted by 'categorizing, explaining and developing arguments that are tied to a specific practical and political problem' (1997: 17). Following this approach, I focus on the relationship between structural injustice and the governance of transformative technologies, AI and Repro-tech, using the UK context as my example. As I shall discuss in later parts of the book, the UK Government boasts that it is both an 'AI superpower' and the world's leader in 'gold standard' Repro-tech governance, thereby providing a useful backdrop against which to rethink governance mechanisms more directly towards the public interest with a structural focus.

CHAPTER 2 ARTIFICIAL INTELLIGENCE AND GROUND TRUTH.

We are the first generation in the history of humanity that has given machines the power to make decisions that historically could only be made by people. If we get it wrong, every generation that follows will pay a price for our mistakes.

Brad Smith, Vice Chairman of Microsoft (2023)

Humanity's increasing reliance on AI and robotics is driven by compelling narratives of efficiency in which the human is a poor substitute for the extraordinary computational power of machine learning, the creative competences of generative AI as well as the speed, accuracy and consistency of automation in so many spheres of human activity. Indeed, AI is fast becoming the core technological foundation of contemporary societies estimated to contribute US$15.7 trillion to the global economy by 2030 (World Economic Forum, 2022).

Most thinking on how to manage the downside risks to humanity of this seismic societal shift to AI and robotics is set out in a direct fault-based relationship. To use a well-known example that caught a great deal of media attention, Elaine Herzberg tragically died when hit by a self-driving Uber car that failed to stop as she wheeled her bike across the road in Tempe, Arizona. Despite being a self-driving car, the Uber car required a safety (human) driver at all times. The 'operator', Rafaela Vasquez on that occasion, had been streaming an episode of the television show *The Voice* and was accused of not watching the road with full attention (Wired, 2022).[8] Was Vasquez guilty of negligent homicide? This is a question of liability (although not necessarily a straightforward one). To take a wider structural perspective, however, is to ask questions such as 'What decisions will be left for humans to make in the future and why does it matter?' Structural questions such as these, I suggest, must be thought of in terms of contemporary politics and governance mechanisms.

Alongside our increasing reliance on AI to think for us, we are also steadily replacing aspects of human labour with robotic alternatives. What interests me here is not so much the usual stories of driverless cars replacing drivers or medical robots replacing doctors, but rather the broader structural dimensions of the fact that the realm of 'jobs for humans' is diminishing.

As Chief Decision Scientist at Google, Cassie Kozyrkov (2020), describes, in AI the truth of an algorithmic calculation is always subject to the desires of its designer and we should not lose sight of this partial capability when transferring more and more human roles over to AI.

Accompanied by a rising public concern about the potential prevalence of unregulated AI-generated harm, states, international organisations and corporations have attempted to address these fears with a range of corporate national and international forms of liability-based governance such as the innovative EU AI Act, which is by far the most comprehensive political attempt to locate (or deter) those directly responsible for AI-generated harm. I argue that while such approaches are vital for combating injustice exacerbated by AI and robotics, too little thought goes into political approaches to the structural dynamics of AI's impact on society. By way of example, I examine the UK's current 'pro-innovation' approach to AI governance and explore how it fails to address the structural injustices inherent in increasing AI usage.

---

[8] Also See BBC (2020).

CHAPTER 3 REPRO-TECH AND THE GENETIC SUPERMARKET. What will our reproductive habits look like in the future, and why does it matter? One part of the answer to this question, for those who can afford it, is Repro-tech. Just before his death in 2015, Carl Djerassi (leader of the team that first developed oral contraception in 1951) made a prediction.[9] He estimated that by 2050 it would be commonplace for women in the wealthier economies to undergo elective egg freezing (or 'proactive' egg freezing) to enable more control over the relationship between their fertility and their careers. Djerassi was convinced that egg freezing would not only lead to profound changes in the reproductive habits of future generations but also raise fundamental questions for humanity itself. By 2019, the technology futurist James Metzl called on his audience of 'high flying professionals' to think about their reproductive choices:

> Raise your hand if you are thinking of having a child more than ten years from now... If your hand is in the air and you are a woman, you should probably freeze your eggs. If your hand is up and you are a man, I encourage you to freeze your sperm as soon as possible. No matter how young and fertile you are ... there's a not insignificant chance you are going to conceive your children in a laboratory, so you may as well freeze your eggs and sperm now when you are at your biological peak. (Metzl, 2019: 1)

Key to both Djerassi's and Metzl's predictions are the rising capabilities of in vitro pre-implantation genetic technologies (PGTs). Originally designed to screen for a range of genetic conditions such as sickle cell disease or Huntington's disease, new markets are set to emerge where prospective parents will be promised the opportunity to select the personality characteristics of their unborn children – what the political theorist Robert Nozick (1974) thought would result in a 'genetic supermarket'. Djerassi predicted that such PGTs would become the 'ultimate factor' in proactive egg-freezing markets, and science fiction would fast

[9] Carl Djerassi was interviewed by Sarah Knapton for *The Telegraph* in 2014. The full article entitled 'Sex will soon be just for fun not babies, says father of the Pill' is available at: www.telegraph.co.uk/news/health/news/11217750/Sex-will-soon-be-just-for-fun-not-babies-says-father-of-the-Pill.html.

become fact with the use of gene-editing technologies. While these technologies would only remain affordable to a minority of the world's population, the structural consequences of genetic enhancement are likely to create an infinitely deeper unequal divide in human society.

Here, the sort of structural perspective discussed in Chapter 1 can help us to think beyond individual usage of IVF, which has brought great positives to human societies, and look to the macro implications of 'fertility insurance markets' and 'genetic supermarkets' propelled by the promise of genetically crafted children at convenient life cycle intervals and the predictions of future fusion of AI and synthetic organisms. Unlike the case of AI, there has been a long-standing tradition of regulating Repro-tech. The UK's Human Fertilisation & Embryology Authority (HFEA) is a regulatory public body created in 1990 in light of a report authored by the philosopher Mary Warnock, and is widely regarded internationally as the gold standard of regulators and the first to govern technologies as complex as gene editing and cloning. However, though we might see some elements of promise in Warnock's approach for a wider model of technology governance, such as an insistence on a much wider scope of professional perspectives than was usually included in technical policy decisions, I chart, nevertheless, what I see as the general demise of regulatory landscapes in line with the dominant US-based 'state capture' school of thought that I discuss in more depth in Chapter 4. In particular, I consider the inherent tensions of a regulator functioning at arm's length from politicians on the one hand but simultaneously being required to facilitate private sector growth on the other. Thinking back to the discussions of Chapter 1, it is also clear that the weight of interest in governance deliberations is largely oriented towards those of the tech industry and its leaders rather than towards more macro-structural considerations that I argue we ought to consider in the context of the public interest.

I suggest such an exploration of the regulatory workings of state governance, which in the UK context employs some 300,000 staff with budgets amounting to over £220 billion, provides some important insights into how we might think differently about the governance of transformative technologies into the future.

CHAPTER 4 PUTTING THE PUBLIC INTO THE PUBLIC BODY. In this chapter, I bring together the arguments from the previous chapters of the book. My reading of Young's work on structural injustice leads us to an uncomfortable political realisation. The usual tools deployed for addressing social harms and injustices – the tracing of liability in contextual moral or legal terms – are not useful for structural injustice, which is much more complex and amorphous in shape, not least in the context of AI and Repro-tech. Young was sceptical that the state could work against structural injustices because of the degree to which states tend to be beholden to the private interests of corporates and other private actors. While I agree on the high incidence of state capture, I argue it need not be so. Indeed, my view is that without the coordinating power of the state to change micro-level behaviour, we have little chance of addressing the negative structural dynamics of AI and Repro-tech despite the extraordinary capabilities of social movements and civil society groups.

Even though no simple political solution is apparent, I argue that one essential approach is to focus specifically on the question of whose interests are at play in the governance of transformative technologies as they operate in the background conditions of structural injustice. Thinking back to the arguments made in Chapter 3 about the functioning of regulators, I argue that the macro-level coordinating powers of a state can be redeployed to address background conditions of structural injustice through the direct reweighting of private and public interest within the mechanisms of governance itself. This is an alternative to current attempts in the context of tech governance to deploy a politics grounded in tracing fault or leaving structural patterns outside political focus altogether. Through a radical reshaping of large-scale regulatory public body landscapes, a new form of lay-centric governance can be incorporated to deliver the sorts of decisions that a state defined by its current relationship with tech industries cannot.

This sort of lay-centric model of public body landscape currently does not exist in governance structures, and an active case needs to be made for its creation.

CHAPTER 5 CONCLUSION: TECHNOLOGY IS THE ANSWER, WHAT WAS THE QUESTION? In my concluding remarks I argue that what I have constructed in this book goes some way to thinking more clearly about how we might address the relationship between transformative technologies, AI and Repro-tech, and structural injustice. That is to say, I have tried to begin give a response to the 'how' of politics rather than only the 'what'.

CHAPTER 1

The Problem of Structural Injustice

Where No One Is Liable, Who Is Responsible?

BEFORE MAKING THE CASE IN THE FOLLOWING CHAPTERS that it is useful to think politically about AI and Repro-tech in the context of structural injustice, I first want to explore the nature of structural injustice itself, why it differs from other sorts of harms and injustices, and what sort of politics is needed to address it.

> The problem with structural injustice is that we cannot trace ... how the actions of one particular individual, or even one particular collective agent, such as a firm, has directly produced harm to other specific individuals. (Young, 2011: 96)

Addressing structural injustice is a seriously difficult political problem. As I shall explore, it's a sort of injustice that, while often closely imbricated with conventional fault-based injustices, cannot be readily addressed in the same way.

Fault-based injustices are always grounded in blame, culpability or guilt, and can be causally traced, whether directly or indirectly, to an identifiable agent or agents (individuals, groups, institutions or states). Addressing fault-based injustices through the tracing of liability in this way is crucial for any legal or moral system to function with a sense of integrity that protects the rights of each individual, group and institution, and requires a minimum standard of behaviour from all. Essential to such a model are clear rules for interpreting evidence of causal connections and establishing intentions, motives and consequences of agents' harmful actions (Young, 2011: 98). However, as I shall detail in this chapter, fault-based solutions are not a productive way of addressing structural injustice.

I begin with Young's seminal work on structural injustice. My interpretation of this work is unusual in that, not only do I defend an element of Young's work, namely a distinction between liability-based injustice and structural injustice, I see political merit in actively sharpening such a distinction. This goes against the grain of current thinking on structural injustice as exemplified by Wolff (2024), who states: 'Now several critics have pointed out that Young's distinction between liability and structural injustice is too stark … This is surely correct.' I hope to convince you otherwise.

WHAT DO I MEAN BY STRUCTURAL INJUSTICE?

There is a large and growing literature on structural injustice and a great many different interpretations as to its meaning and impact.[1] My interpretation of structural injustice is that it is characterised by negative outcomes experienced by individuals, groups or institutions that are generated by a vast array of complex, multitudinous 'everyday' actions of those and other individuals, groups and institutions operating within accepted norms and institutional practices in a given time and space and accumulated at the macro level (Young, 2011: 53).[2] These outcomes, which are inherently intersectional,[3] serve to reinforce how individuals, groups and institutions are positioned in relation to each other in ways that shape the opportunities and life prospects (including both material and social resources) of everyone in those positions and their future actions (Young, 2003: 6).[4] We might think of these structural relations

---

[1] See, for example, Gordon (2007), Nussbaum (2011), Parekh (2011), Haslanger (2015), Barry and MacDonald (2016), Brooke (2018), Goodhart (2018), Corwin and Jaggar (2018), Lu (2018), Sangiovanni (2018), McKeown (2018), Aragon and Jaggar (2018), Zheng (2018), Nuti (2019), Powers and Faden (2019), Atenasio (2019), Jugov and Ypi (2019), Beck (2020) and Chung (2021).

[2] We might think of this as a 'wrong that is created by no wrongdoing'.

[3] See, for example, Zheng (2018) and Hill Collins and Bilge (2020), whose work serves to demonstrate how the various strands of structural processes that affect particular gendered or racialised groups and individuals are co-constituted and serve to mutually reinforce each other.

[4] Here Young acknowledges the work of many others such as Bourdieu (1984), Kutz (2000), Sartre (1976) and Sewell (2005).

and practices as 'the background conditions' of structural injustice or as a vast structural matrix which, as Hill Collins explains, 'is not a benign container in which something happens, but rather shapes and gives structure to dynamic phenomena' (2017: 24). Through this immensely complex structural matrix, individuals, groups and institutions are connected to a multitude of structural injustices without being meaningfully traceable to them.

Rather than assume that absolutely everything we do is connected to every structural injustice, for clarity's sake I will name the habits, expressed beliefs and actions that we might speculate are infused within the background conditions of a particular structural injustice as 'structural actions'. In other words, we might conceive of structural acts as among 'legitimate' pursuits of private interest in a given time and place, in contrast to the legal and moral wrongs that are causally traceable to fault-based injustice.[5]

This distinction is a much stricter interpretation of Young's account than previous thinkers on structural injustice. Why do I think this distinction is important? As I hope will become clear in the examples and discussions that follow, if we restrict ourselves to traceable fault-based solutions to untraceable structural problems, we will not succeed in achieving effective progressive change. Rather, if we think of structural

---

[5] This idea is in some sense the reverse of Adam Smith's 'invisible hand' interpretation of market forces, whereby the decentralised individual actions of private vice (greed or selfishness) unintentionally generate what Smith thought of as a *public virtue* in the form of an efficient market (2010 [1759]: 161). However conversely, the decentralised individual actions of private virtue (legitimate pursuits of private interests) unintentionally generate a *public vice* in the form of structural injustice (thanks to Jason Sharman for pointing this out). Indeed, on this particular point, Young's perspective was closer to that of Max Weber, who lacked faith in 'private vices' to spontaneously morph into 'public virtues' and social order, often because of the negative unintended consequences that they tended to produce (see, e.g., Turner, 2019: 385). For Weber, social order required the state and law, and his consciously imperfect solution was the 'berufspolitiker', the professional politician of the modern state, who, if endowed with the right vocation, would be able to counter an ethic of conviction (moral passion) with that of an ethic of responsibility (a sensibility to consequences) so as to choreograph social order. As we shall see, Young thought the answer lay not with politicians but with civil society. While acknowledging the importance of her approach, this book takes a somewhat different tack.

injustice and its causes as a different sort of injustice from that which is generated by liability (legal and/or moral), then we have a better chance of enabling a more promising political response.

Those familiar with Young's work will know that in order to demonstrate how structural injustice operates on her account, Young constructed the 'story of Sandy' as an analytical device designed to isolate the particular ways in which structural injustices emerge. Sandy is a single mother who sets out to find a rental flat within a reasonable commuting distance from her city-central place of work and her children's school. While the extent of Sandy's story is not recounted in full here,[6] suffice it to say that due to a commonplace set of issues surrounding access to decent affordable housing (involving the preferences of house-buyers and property developers, high costs of commuting, expensive childcare services, rise in rents, stagnation of wages and declining employment opportunities to name but a few), Sandy and her children find themselves in dire circumstances despite their best efforts and through no fault of their own or any other particular agent.[7] Young argued that, although 'no particular agent she encounters has done her a specific wrong' (2011: 47), 'Sandy's story illustrates a specific kind of . . . wrong, structural injustice, which is distinct from wrongs traceable to specific individual actions or policies' (2011: 44). Sandy stands in relation to hundreds of thousands of other people who are apartment-renters, house-buyers, landlords, employers, commuters, etc., all in pursuit of their private interests as part of a vast structural matrix of social positioning. In real life, some of these agents may well be blameworthy (unscrupulous landlords, bad government policies), but the point that Young was trying to make is that many of the agents and institutions in Sandy's story will not be at fault (such as those following better labour opportunities or perhaps featuring in

---

[6] See Young (2011).

[7] Here I am reminded of Michelle Obama's (2020) speech to US graduates in 2020:

> The truth is, when it comes to all those tidy stories of hard work and self-determination that we like to tell ourselves about America, well, the reality is a lot more complicated than that. Because for too many people in this country, no matter how hard they work, there are structural barriers working against them that just make the road longer and rockier. And sometimes it's almost impossible to move upward at all. (Obama, 2020)

international migration patterns and so on), and yet their actions should be taken into account when thinking through political solutions.

I see this as a profoundly political argument in that to distinguish the two forms of injustice (fault-based and structural) is to argue that structural injustice should be properly identified as *injustice* that requires a political response just as fault-based injustice does. All too often, structural injustice is overlooked as simply the outcome of misfortunes experienced by certain individuals, groups or institutions.[8] However, '[t]o judge a circumstance unjust implies that we understand it at least partly as humanly caused, and entails the claim that something should be done to rectify it' (Young, 2011: 95). In other words, a political approach is required. Even though structural processes only exist as a consequence of individual, group or institutional behaviour, they are, nevertheless, experienced as constraining 'objective social facts' (Young, 2011: 53), which in the case of Sandy is the material position of not being able to afford a home for herself and her children through no fault of her own and is not a dynamic she can address on an individual level.

Despite its simplicity, Young's story of Sandy has generated many different interpretations.[9] For me, the most important implication is that asking who is to blame for Sandy's plight is not a question that captures the structural injustices she faces: 'it's not possible to trace which specific actions of which specific agents cause which specific parts of the structural processes of their outcomes' (Young, 2003: 7).[10] One might conclude from this, as many of Young's critics do, that those whose entrenched privilege is protected by the law and moral norms of the day are seemingly 'off the hook' for structural injustice. However, I see the importance of Young's account lying in quite the opposite interpretation in that it requires structural injustice to be collectively addressed *despite* operating beyond the ability to trace agents of fault. As we shall see in the following chapters, this is the key point for thinking about political approaches to structural dynamics of AI and Repro-tech.

The idea that structural injustice ought not to be understood in terms of liability was inspired by Young's critical engagement with the work of

---

[8] As I shall discuss later, this is Reiman's argument (2012), for example.
[9] See, for example, Browne and McKeown (2024).   [10] Also see Young, 2011: 109.

Hannah Arendt. Writing in 1945 in the context of the Third Reich's orchestrated genocide of Jewish people, the question of whether the German population and beyond should be identified as 'guilty' for the rise of the Nazis in Germany and their subsequent acts of genocide was a central concern for Arendt: 'Guilt, unlike responsibility, always singles out; it is strictly personal' (2003 [1968]: 147). Arendt had argued that a charge of guilt should not be diluted by including those who collectively failed to act against injustice[11] but who, nevertheless, did not commit a heinous crime: 'Where all are guilty, nobody in the last analysis can be judged' (1994 [1945]: 126).[12] Those who are prepared to act in such a way that harms others should be held to account. Their guilt ought not to be watered down by the inclusion of those at the periphery of wrong-doing.[13] For Arendt, then, the discovery and consequent assertion of guilt is a backward-looking exercise in identifying an individual, group or institution for their morally or legally wrongful actions[14] (these include

---

[11] As Young (2011: 90) discusses, Arendt recognises that many of those who did not act against injustice refrained because they were living with the everyday acts of brutality and murder of the Nazi state. Indeed, Arendt's only example of political responsibility was that of the Scholl siblings, who were founder members of the White Rose Movement, a non-violent anti-Nazi student group during the Third Reich. Both Sophie and Hans Scholl were tried and executed for publicly inciting resistance against the Nazi's persecution of the Jews and Poles as well as other SS atrocities. See Arendt (2006 [1963]: 104) and also Scholl (1983).

[12] This statement featured several times in Arendt's work (see, e.g., 2003 [1968]: 21, 28, 147).

[13] My intention here is not to verify or dispute Young's interpretation of Arendt's writings but rather to develop my own interpretation of Young's thought on political responsibility, which in part is based on her reading of Arendt's writings. Here my intention is to sharpen the concepts of structural injustice and political responsibility and take them in a different direction from that of Young in the latter parts of this book. It is also important to note that I am not suggesting that my work is relevant to the context of the Holocaust, which was the focus of Arendt's work cited in this book. Furthermore, Arendt's work is deeply problematic on questions of identity (notably gender and race). Somewhat ironically, nevertheless, her theoretical work on what it means to take up political responsibility to facilitate the plurality of humanity serves to illuminate the potential structural threats of technology, as I shall discuss further on in the book.

[14] As Arendt explains, '[l]egal and moral issues are by no means the same, but they have a certain affinity with each other because they both presuppose the power of judgement' (2003 [1964]: 22).

being liable in virtue of holding particular office or by not acting when one is in a position to prevent harm). This for Young is an exercise in 'the liability model of responsibility for injustice',[15] and for both theorists, such an approach requires the tracing of guilt as a necessary prerequisite for devising suitable sanctions and other political responses.

For Young, the alternative to the liability model of responsibility for injustice is 'political responsibility', and this too was an idea she recovered from the work of Arendt. For both theorists, political responsibility is characterised by collective public political action aimed at injustice that is distinct from the legal responsibility (required observance of the law) and moral responsibility (abidance to expected or prescribed social norms) in a given context. To act against injustice collectively is of course a moral responsibility of sorts,[16] in that it is about framing and addressing injustice, but not in the sense of 'moral fault' as prescribed by the moral codes of a given place and time that create expectations of each individual and which, if transgressed or neglected, bring about the charge of guilt. As Arendt's account (2006 [1963]) of the Third Reich reminds us, there are stark examples in human history where abidance by the law or the moral codes of the day have served grave injustice. Neither, though, for Arendt, is political responsibility an individual moral act of personal conscience (which may well bravely clash with

---

[15] Here the wording becomes a little confusing. In *Collective Responsibility*, Arendt argues, 'There is such a thing as responsibility for things one has not done; one can be held liable for them. But there is no such thing as being or feeling guilty for things that happened without oneself actively participating in them' (2003 [1968]: 147). Young chooses the word 'liability' to capture all those cases where one is traceably responsible for an injustice. This includes any fault-based responsibility for an act of wrongdoing, whether it be directly, indirectly, intentionally or unintentionally as well as being liable in virtue of holding office (what Young calls a 'bureaucratic chain of command' (2011: 115)). Young argues that 'the liability model of responsibility ... is inappropriate for assigning responsibility in relation to structural injustice' (2011: 98). While this definition is not always consistent in *Responsibility for Justice*, I find this particular account of the liability model convincing, and build upon it to develop my account.

[16] As Young explains, 'in ordinary language we use the term "responsibility" in several ways ... In this [social connection model], finding responsible does not imply finding at fault or liable for a past wrong; rather, it refers to agents' [*sic*] carrying out activities in a morally appropriate way and seeing to it that certain outcomes obtain' (2011: 104).

the moral norms of the day).[17] Rather, it is a collective stock-taking of moral and legal norms and institutional behaviours combined with the willingness to act with others to address injustice where injustice is found. As Arendt argues, '[i]n the centre of moral considerations of human conduct stands the self [whereas] in the centre of political considerations stands the world' (Arendt, 2003 [1968]: 153; Young, 2011: 89).

Arendt's separation of guilt from political responsibility was an idea that Young sought to contend with in a very different sort of context from that which had occupied Arendt, however. Rather than the extraordinary horrors of the Holocaust, Young looked to what she thought of as the 'ordinary' structural injustices of everyday life that were often intertwined with increasing economic precarity under conditions of neoliberalism,[18] institutional racism, sexism and what is referred to today as exclusionary populist politics (Moffitt, 2020). Consequently, instead of relying on the establishment of guilt, fault or liability to generate a political response to structural injustice, Young began to articulate a supplementary account of a different sort of discretionary collective political responsibility, drawing on Arendt and focused on injustices that could not be meaningfully traced as the fault of any agent in particular: 'when the injustice is structural, there is no clear culprit to blame and therefore no agent clearly liable for rectification' (Young, 2011: 95). In this sense, Arendt's observation that '[w]here all are guilty, nobody ... can be judged' is turned on its head to ask *where nobody is liable, who is responsible?* This question acts as a central theme of this book and will serve as a way to explore structural aspects of tech governance in subsequent chapters.

For both Young and Arendt, political responsibility is something to be taken up: it is the responsibility *to be political.* However, the question of what motivates political responsibility presents an informative point of

---

[17] Arendt gave examples of many individuals whose moral acts were extraordinarily brave under the Third Reich, such as those who hid Jewish people in their homes or refused to swear allegiance to Hitler (Arendt, 2006 [1963]: 104). The story of Austrian farmer Franz Jägerstätter, who refused to serve in the German Wehrmacht, is another sobering example. Arendt's view on this point was Socratic: 'it is better to be at odds with the whole world than, being one, to be at odds with myself' (2003 [1968]: 153).

[18] See for example Brown (2019).

departure by Young from Arendt: Young saw political responsibility as the appropriate response to structural injustice in the context of 'ordinary injustice', and for her this meant moving beyond the motivations of a distinct political community (such as the state) that characterised Arendt's account.[19] Arendt had assumed that political responsibility must be motivated by political membership; 'a collective and vicarious responsibility in which the member of a community is held responsible for things he did not participate in but which were done in his name' (2003 [1968]: 154). For Arendt, guilt, in line with its opposite concept, innocence, is directly linked to a particular agent, while political responsibility could be vicarious in virtue of one's membership of a political community (29). In this sense, the state is also subject to political responsibility 'which every government assumes for the deeds and misdeeds of its predecessor and every nation for the deeds and misdeeds of the past' (27). For Young, however, Arendt's argument did not make sense. Rather, she argued that *action* is the motivating factor of political responsibility; 'it is a mystification to say that people bear responsibility simply because they are members of a political community, and not because of anything at all that they have done or not done' (Young, 2011: 80).[20] On Young's view, everyone (wherever they are and however they identify) who *contributes to structural injustice is responsible*, not by having personally caused or intended injustice, but because they have acted in ways that have contributed to the background structural conditions of structural injustice through their legitimate pursuits in everyday life. Their collective 'social connection' is the motivation for political responsibility, hence Young's name for the alternative approach to the liability model of responsibility – the 'social connection model'.

This departure from Arendt also leads Young to conclude that the state[21] is not the most productive mechanism to generate political responsibility for structural injustice. In fact, Young argues that often:

[19] See for example Arendt (2003 [1968]: 149).

[20] This is especially so as Arendt's understanding of politics is grounded in speech and action epitomising active life – what she called 'viva activa'. See Arendt (1998 [1958]: 7–17).

[21] Young provides a definition of her understanding of the terms 'state', 'economy' and 'civil society', which work equally well for the focus of this book: 'State refers to activities

rules and practices of [state] institutions are more aligned with the powers and processes that produce or perpetuate injustice than with those who seek to undermine it. We cannot turn to the state or international institutions as arbiters in a struggle between the interests that produce structural processes with unjust outcomes and interests in changing those processes. The policies and programs that states and international organizations enact themselves tend more to reflect the outcome of those struggles than to balance between them or adjudicate them. (2011: 151)

This leads Young to weight her conception of political responsibility for structural injustice towards civil society and away from the state. Within the structural injustice literature, there is a strong tendency to take the same view in the face of state failures to address structural injustices (see, e.g., Schiff, 2014; Powers and Faden, 2019; Haslanger, 2022). I shall take a different approach to political responsibility and its relationship to the state later in this book, although not for the same reasons as Arendt. It is nevertheless illuminating to appreciate Young's aim of carving out an interpretation of political responsibility that was not bound to the backward-facing exercise of establishing guilt like Arendt's. Rather, and drawing on scholars such as Goodin (1998),[22] Young wanted to cast political responsibility for structural injustice as more of a collective, forward-looking, shared endeavour to effect change. I have some reservations about the absolute distinction between these two forms of responsibility which I set out towards the end of this chapter, but first I want to

and institutions of legal regulation, enforcement backed by coercion, legislatively mandated co-ordination and public services, along with the managerial and technical apparatus necessary to carry out these functions effectively. In distinguishing economy from state, I assume a capitalist economy, that is, an economy in which at least a large part of the society's goods and services are supplied by private enterprise operating through markets. Economic activity is profit- and market-oriented. Civil society refers to a third sector of private associations that are relatively autonomous from both state and economy. They are voluntary in the sense that they are neither mandated or run by state institutions, but spring from the everyday lives and activities of communities interest' (2000: 158).

[22] Collective responsibility 'is forward-looking, task oriented' (Goodin, 1998: 50). Also see Schwenbecher (2010), Pinkert (2014) and Pettit (2007). These authors differ from Young, however, in that they all hold liability as central to their accounts. Arendt on the other hand did not think that collective responsibility tallies with guilt; see Arendt (2003 [1968]: 147).

explore and refine that which I defend and sharpen in Young's account as well as highlight the elements of Arendt's work that help to clarify political responsibility.

THOUGHTLESSNESS AND STRUCTURAL SPECULATION

In order to better capture what is meant by political responsibility, I find it illuminating to think of what might be its opposite. A concept that Young raises from Arendt's work, only in passing,[23] is that of 'thoughtlessness'. For me however, 'thoughtlessness' tells us something fundamental about the failures of political responsibility.

Arendt described a form of 'thoughtlessness' which in some of her work can be interpreted as explaining how law-abiding people fail to take up their political responsibility.[24] '[T]houghtlessness – the heedless recklessness or hopeless confusion or complacent repetition of "truths"

[23] See Young (2011: 84, 87) where 'thoughtless' is mentioned in passing by way of describing Arendt's analysis of the Eichmann's trial (Arendt, 2006 [1963]). While Arendt thought Eichmann was clearly guilty for his part in the Holocaust, she also warns us to recognise the dangers of a political system that routinises behaviour in such a way that independent reflection is diminished. Arendt's point was not to excuse individuals such as Eichmann but to suggest (as she had done previously in the prologue of *Human Condition* (1998 [1958])) that such thoughtlessness was prevalent throughout society and in some cases reflected her famous phrase, 'the banality of evil' (Arendt, 2006 [1963]: 252).

[24] See Benhabib (2018: 70–79) for an in-depth discussion of Arendt's concept of thoughtlessness in the context of Adolf Eichmann's trial and the differences between Arendt's ideas on 'thinking' and those of Heidegger. Benhabib provides a compelling account of how Arendt may well have misconceived Eichmann's personality and the extent of his anti-Semitic prejudices despite finding him guilty of heinous crimes. Nevertheless, the conceptual idea of thoughtlessness which is most fully developed in Arendt's account of Eichmann's trial remains altogether valuable in explaining how people lose, or perhaps do not develop, the capacity to genuinely 'think' critically about what they are doing and its effects. In both Seyla Benhabib's (2021: 276) and Judith Butler's (2011: 280) reading of Arendt, thoughtlessness is interpreted as 'obedience', meaning that people continue to orientate their actions towards required institutional goals irrespective of the harm that it does to others. Certainly I find it hard to believe that Eichmann did not understand himself to be committing acts of atrocity, no matter how entrenched in bureaucratic norms he found himself to be. However, the idea of thoughtlessness as a habit of inertia so deeply ingrained it does not occur to people that there is a possible or viable alternative is a powerful one, and it is this sort of habit I mean by 'thoughtlessness' in this work. See Schiff (2012, 2014) for further interpretations of the different ways in which Arendt's concept of thoughtlessness operates in her work.

which have become trivial and empty – seems to me to be among the outstanding characteristics of our time' (Arendt, 1998 [1958]: 5). One way of reading this is that thoughtlessness is not absent-mindedness but rather a sort of habitual disengagement from thinking critically about the plight of those negatively affected by a community's institutions and collective behaviour. For Arendt, the extraordinary achievement of the Nazis during the Second World War was to capitalise on an 'international modern phenomenon' whereby 'under pressure of the chaotic economic conditions of our time', an ordinary individual would transform from being 'a responsible member of society, interested in all public affairs, to a "bourgeois" concerned only with his private existence' under the Third Reich (1994 [1945]: 129). This phenomenon resulted in 'the vast machine of administrative mass murder' which employed, routinised and normalised the behaviour of so many ordinary German citizens who in other circumstances would never have contemplated contributing to human suffering (1994 [1945]: 126). This thoughtlessness was an evacuation of the realm of political responsibility. Indeed, thoughtlessness is an affront on politics itself.[25] To 'think'[26] brings with it the potential to formulate new political judgement, 'to tell right from wrong' and to devise political action (Arendt, 1971: 418). This sort of judgement does not require special forms of expertise or training but rather is a habit of reflection that manifests when in the company of others and exposed to their different thoughts and actions (1971: 446). Thinking has the capacity to construct new possibilities beyond existing legal and moral norms and rules – that is to say, the capacity for original political judgement and social transformation.[27] In this sense, what is required is to work against thoughtlessness by enabling the flourishing of pluralistic thought and with it, the possibility of politics itself. As Arendt argues, this is not an exercise in following 'doctrines' but rather looking to the 'experiences'

---

[25] As Schiff explains: 'Plurality is not just about multiplicity, about the coexistence of many rather than the existence of just one. It is more fundamentally about connections between us, connections that make it intelligible to say that we share the world in common' (2014: 54).

[26] 'What I propose therefore, is very simple: it is nothing more than to think what we are doing' (Arendt, 1998 [1958]: 5).

[27] See Steinberger (1990) for a fuller discussion.

of the 'everybody' (1971: 71). Here we might think of the public interest being determined not as a fixed idea (that would be to follow doctrine) but rather a more fluid politics based on the possibilities of new thinking derived from the everyday in a given time and place.[28]

For Arendt, the plurality of humanity offers endless such political possibilities:

> [T]he character of startling unexpectedness is inherent in all beginnings ... The fact that man is capable of action means that the unexpected can be expected from him, that he is able to perform what is infinitely improbable. And this again is possible only because each man is unique, so that with each birth something uniquely new comes into the world. (1998 [1958]: 178)

Each of us brings the potential for a new way of understanding and acting.[29] However, this is an argument not for increased populations per se but rather for the protection and enabling of human diversity in an environment where critical thinking for oneself is considered vital.[30]

Contemplating the context of 'ordinary injustice' characterised by increasing social and economic precarity, Young did not make explicit use of Arendt's idea of 'thoughtlessness' for her account of structural injustice.[31] However, under the pressures of everyday life, 'thoughtlessness' includes an unquestioning habitual abidance to norms and social rhythms which generate the habits and expressed beliefs that collectively, and in large part, constitute the background conditions of a particular structural injustice. Although I find Arendt's concept of thoughtlessness valuable, structural actions are conceptually distinct from the fault-based wilful blindness to injustice that Arendt sought to highlight as her primary focus (such as the case of Eichmann).[32] My meaning of structural

---

[28] Here Arendt is particularly influenced by the Socratic approach to politics (Arendt, 2005 [1954]: 6).

[29] '[B]ecause we are all the same, that is, human, in such a way that nobody is ever the same as anyone else who ever lived, lives, or will live' (Arendt, 1998 [1958]: 8).

[30] See Benhabib (2018) for a fuller discussion of this point and its influences.

[31] See Footnote 23.

[32] Arendt of course was not writing about a time of 'ordinary injustice', and her intention was to describe the ways in which some otherwise decent people lived and worked under an abhorrent regime. There is no doubt that some of the thoughtlessness she describes is

actions is such that it refers to those thoughtless actions that operate within the realm of everyday legitimate pursuits of private interest in a given time and space.

Taking up political responsibility to address structural injustice requires us to 'think' in order to collectively act and effect change, but of course the context in which we do this is all important. I shall consider this challenge in later parts of the book with a particular focus on tech governance.

Determining what we, as individuals, groups and institutions, are responsible for as a consequence of relational positioning within a structural matrix requires the development of a political habit of thinking about the dynamics of the social structures in which we are embedded and the ways in which our structural actions both affect and are affected by those structures. In Young's account of structural injustice, she describes a 'structural point of view' (Young, 2011: 70) as a way of seeing the harmful consequences of 'patterns in relations among people and the positions they occupy relative to one another ... [and] to see how the actions of masses of people within a large number of institutions converge in their effects to produce such patterns and positioning' (2011: 70–71). However, even within the cases of seemingly straightforward fault-based injustice, establishing causality between an agent and an injustice can be difficult – perhaps there is only a correlation between factors or an undetected confounding variable that is the real cause behind a correlation between two occurrences (Beebee et al., 2012). Nevertheless, conventions around evidence have emerged, whether through our faith in scientific method or our willingness to accept consistent correlations as 'good enough' proof of causation. Certainly, these are the tools of tort law, and by and large they provide a vital function of society which operates on varying belief systems expressed through shifting moral and legal rules. However, as I have set out, contributions to the background conditions of structural injustice are too convoluted to be meaningfully traced in the sense of liability to any particular agent, and for that reason it is not productive to ground political responses in tracing connections between structural injustices and

wilful blindness to injustice, but some of it will also be closer to my concept of structural action.

particular individuals, groups and institutions. And yet, there needs to be a way for individuals, groups and institutions to take up a structural point of view on their contributions to the background conditions of structural injustice. This seems an intractable problem. What I suggest we need is more active speculation or more precisely 'structural speculation', akin to Arendt's 'thinking' habit. However, as I shall explain, such speculation does not seek to determine right from wrong or good from evil as Arendt's 'thinking' does, but rather focuses on which actions might conceivably be structural actions contributing to the background conditions of structural injustice and which could be the basis of taking up a structural perspective and becoming politically responsible for effecting change.

The idea of 'speculation' in traditional legal parlance is predominantly negative. It is understood as opinion, hypothesis or conjecture about what might happen in the future rather than the representation of a proven fact. It is the basis on which claims are rejected in virtue of the lack of hard evidence. It is insufficiently traceable to any concrete verification of liability. As legal scholars Hillel et al. (2005: 229) demonstrate, causation or liability 'must be rooted in the evidence and not the product of mere speculation'. A brief glance through case law gives many such examples: 'It is not for the Judge to speculate in the absence of evidence'; 'the Judge's speculative approach arguably contaminated his ... analysis'. However, I see the active development of a habit of structural speculation – thinking about which of our habits, beliefs and actions may be structural actions – as a necessary precursor to developing a collective structural *point of view* and acting to effect change as part of political responsibility.

How to translate this rather abstract objective into the practicality of politics is a question I return to later but suffice it to say here that throughout the book I refer to speculation as the counter-concept to thoughtless structural actions which as I shall argue further on is particularly important when thinking about the structural dynamics of society's increasing reliance on AI.

While I have thus far sharpened Young's distinction between liability and structural injustice, I am not convinced by all of the elements of Young's account of political responsibility. However, before advancing that claim, I want to demonstrate in a little more detail why a liability-based politics is not sufficient for grounding responses to structural injustice. I shall do this by critically engaging with the recent work of several advocates of liability-based political action (many of them recent critics of Young's work). This analysis will serve to underpin a central argument of the book – that politics and its attendant policies (not least those focused on the governance of transformative technologies) are too heavily weighted towards a liability-based approach and insufficiently focused on the structural.

THE LIMITS OF LIABILITY

'[R]esponsibility in relation to structural injustice should not be thought of as an attenuated form of responsibility as complicity . . . What we should seek is not variation on a weaker form of liability but rather a different conception of responsibility altogether.

Young (2011: 103–104)

Nussbaum (2011), Barry and MacDonald (2016), Goodhart (2018), Sangiovanni (2017, 2018), Powers and Faden (2019), Atenasio (2019) and many others are not convinced by Young's claim. Their view is that Young's non-liability approach to structural injustice 'absolves' culprits, letting 'the slate be wiped clean' (Barry and Ferracioli, 2013: 256) in not performing retrospective accounting of discharged responsibilities – what Goodin calls 'a shadow of the past' (1998: 150).

Powers and Faden, for example, explicitly 'part company' with Young on the 'neatness' of liability-based injustice:

We do not share her assumption that we can isolate those cases in which the origins of structural injustice are relatively benign from morally more tainted cases . . . [and which] do not come ready-made in neat analytic categories with exploitation, subordination and exclusion appearing in one scenario and largely benign causal origins of structural injustices in another. (2019: 114)

Accordingly, Powers and Faden include human trafficking, police misconduct and voter suppression tactics as 'all too familiar examples' of structural injustice (2019: 1). It is certainly right that these sorts of injustices will inevitably be imbricated with millions of structural actions that contribute to increasing economic precarity, which in turn, drives individuals into exploitative and desperate situations. Whilst it may be difficult to disentangle the multiple aspects of both liability-based and structural injustice, the act of human trafficking is not an untraceable structural acts within legitimate pursuits of private interest. Contra Powers and Faden's critique of Young, there is, in fact, nothing 'neat' about the complex interwoven aspects of an injustice – some fault-based, some structural – but it is imperative that those aspects which are clearly wrongdoings need to be, as Arendt put it, 'singled out'[33] and the culprits ought to be identified with those actions for which they are guilty. Ironically then, Powers and Faden's account serves to 'neaten up' the complexity of injustice by bundling structural injustice together with guilt-based injustice under liability. The untraceability of structural injustice, that Powers and Faden deny, is precisely why, on the understanding of structural injustice I have set out, there needs to be a different kind of political approach to addressing it.

In a similar vein to Powers and Faden, Barry and MacDonald (2016: 98) propose that Young's account of the liability model is 'unduly narrow' and that her interpretation of 'causal connection' is 'too restrictive'. They demonstrate their point with an illustrative example:

> The fact that some agent's contribution to some harm is not necessary to its occurrence – as when 10 people push a rock down a hill where it crushes a car when any 5 of them would have succeeded in generating enough force to do so – does not mean that they are not liable to bear the cost required to address it. When many people make non-necessary but jointly sufficient contributions to harm, they can be held jointly and severally liable for it. (98)

This is of course correct, but Young is clear that her definition of liability would not only encompass all ten rock-pushers but is in fact a more comprehensive and 'strict' account of liability than Barry and MacDonald's example:

---

[33] As I described previously, Arendt argues: 'Guilt, unlike responsibility, always singles out' (2003 [1968]: 147).

> Under strict liability, the law holds an agent liable for a harm even if the agent was not the cause of the harm and did not intend or was unable to control the outcome, such as when one person's property causes damage to another person's property ... Under what I call the liability model of responsibility, I include all such practices of assigning responsibility under the law and in moral judgement that seek to identify liable parties. (Young, 2011: 98)

My argument here is that the act of trafficking another human being, pushing a rock with the intention of crushing a car or, as I suggested in the Introduction, creating an algorithm that discriminates against a particular group of people is not a 'structural action' but rather a wrongful act captured by the liability model. Despite the imbrication of different forms of injustice, liability-based and structural, they are nevertheless conceptually distinct and this is important because each requires a different political approach.

My particular interpretation of untraceable structural injustice puts my account in tension with the work of my co-author, Maeve McKeown, with whom I edited the book *What is Structural Injustice?* and it is helpful for my argument in later parts of this book to take some time here to explain why.[34] McKeown (2024) has constructed a compelling typology of structural injustice through a critique of another of Young's famous structural injustice examples – that of the anti-sweatshop movement. Young explains how this movement has achieved some success in its attempts to bring awareness to consumers through new forms of social media and communication, illicit responsive action from global fashion and sports brands as well as corresponding policy change by producers. Thanks to anti-sweatshop activists (such as the United Students Against Sweatshops)[35] operating across different national and legal jurisdictions

---

[34] See Browne and McKeown (2024). This collaboration was incredibly creative and a fantastic experience of how productive and constructive intellectual disagreement can be.

[35] See United Students Against Sweatshops website at the following link: https://usas.org Here you will see campaigns like the following; 'Nike's supplier Hong Seng Knitting factory used threats and intimidation to steal a total of $600,000 from garment workers who sew our university apparel. USASers across the country refuse to allow our

over the past twenty years, there have been some recorded reductions in the very worst aspects of sweatshop labour in a range of contexts, including the use of child labour, the eighty-hour working week and the use of petroleum-based adhesives in the production process that cause damage to the liver, kidney and central nervous systems of factory workers in apparel and footwear factories used by brands such as Nike (see Williams, 2020). Young describes how civil society groups have managed to draw in huge numbers of individuals from all over the world to their campaign and helped them to develop a structural perspective on the complex relationship of international consumption patterns to specific labour conditions of production. Through these civil society efforts, many more of us have been made aware that to buy particular brands of cheap clothing in local stores or online contributes to some structurally pernicious sweatshop conditions of workers elsewhere. In response, McKeown (2024) argues there are, in fact, a great many agents involved in international consumption patterns that ought to be held to account for their part in creating exploitative working conditions, and that Young's example does not pay enough attention to the power dynamics and private interests at play.

For McKeown, the defining feature of structural injustice is agency rather than structural actions or untraceability, and, unlike my account, she incorporates mal-intent and wilful negligence into her definition of certain forms of structural injustice. 'Pure structural injustice' is cast as the 'unintended, unforeseeable' injustice for which 'there are no agents with the capacity to remedy it' (2024: 4). 'Avoidable structural injustice' on the other hand, is foreseeable and results from the actions of powerful agents with the capacity to remedy the injustice, but who nevertheless do not. Finally, '[d]eliberate structural injustice' is purposefully choreographed by powerful agents precisely because they benefit from it. McKeown describes how international organisations, international business lobbies, a network of global financial institutions and consumers all form a 'social alignment that maintains the power of multinational

universities – who hold multi-million dollar contracts with Nike – to be complicit in mass wage theft. Take action now on your campus, in your community, and online'.

companies (MNCs) over sweatshop workers' (81). As McKeown explains, 'there are measures that any one of these groups could take to improve workers' rights but, for the most part, they do not (81).

McKeown's account is intuitively persuasive. However, from the perspective I have set out here with an emphasis on untraceability, McKeown's descriptions of both avoidable and deliberate forms of structural injustice would be captured by the category of liability-based injustice. For example, on my account, McKeown's case of the '[m]ultinational corporations (MNCs) in the global garment industry [who] deliberately perpetuate sweatshop labour ... through lax codes of corporate social responsibility and manipulation of consumers' (81) does not describe a structural injustice but rather a traceable and morally liable one. I suggest that the sorts of negligent or intended action McKeown describes do not fit with the idea of structural action, and that structural injustices have a much more convoluted relationship with particular individuals, groups and institutions than set out in McKeown's account. Like many of Young's critics, McKeown insists on relocating structural injustice inside the liability-model of responsibility with her categories of 'avoidable' and 'deliberate forms of structural injustice', and argues that it is 'questionable whether there are any cases of pure structural injustice' at all (82). In one sense, I agree with McKeown that the way in which Young describes the processes that generate sweatshop labour in her 2011 work do appear to suggest malintent and negligence, and this does not fit with Young's own original argument of distinct types of injustice. However, following my stricter interpretation of that distinction, I suggest that such an example serves to demonstrate a vastly complicated combination of both fault-based and structural injustices closely operating together, each requiring a different sort of political approach nevertheless and resulting in a wider set of issues being brought under political scrutiny.

From an alternative critical angle, Sangiovanni (2017, 2018) suggests that Young's account of structural injustice and, more specifically, what I am calling 'structural actions', could in fact be interpreted as unintentional acts of indirect discrimination which would bring them inside the liability model. Sangiovanni argues that if individuals are causally linked to injustice, no matter how diffuse or unaware, then they must be liable for that injustice.[36] By way of example, Sangiovanni presents an amended version of Parfit's famous 'harmless torturer' (1984) to make his point:

> Suppose that someone is wired to a torture machine connected to fifty million switches. If none of the switches is flipped, then no current runs through the machine. If fifty million switches are flipped, the person will be in severe pain. But no one switch makes a discernible difference to the pain experienced by the man connected to the machine, no matter how many switches have been flipped previously. It is only the combined effect of the switches that makes a difference. Imagine that each switch is operated by a single individual, and that each person knows that, by flipping the switch, they contribute (infinitesimally) to the torture. And also assume that the switches are also connected to each person's house lights, so there is no way for them to turn on their house lights without sending the charge through to the tortured man, and no way to avoid doing so without collectively organizing to rewire the electricity network. (2017: 165; 2018: 469)

Sangiovanni sees the dynamics of this example as reflecting those of Young's social connection model of political responsibility in that the perpetrators of injustice do not see themselves as significant contributors, and so do not deem it imperative to change their particular behaviour (2018: 166). Sangiovanni's claim is that '[e]ven if people do not know exactly how their small-scale actions, decisions, and patterns have such effects ... and that there are reasonable alternatives, their indifference becomes (at least *pro tanto*) objectionable' (2017: 168).

Assuming that there are 'reasonable alternatives' for people to act differently, on this view Young's account of structural injustice fails to escape the fault-orientated liability model.[37] However, I think Sangiovanni misconstrues two important elements here and if I am right, we can hold that

---

[36] Andrea Sangiovanni and I have had several fascinating conversations about this point (see Sangiovanni, 2017: 168). His book *Humanity without Dignity* has presented me with perhaps the most difficult challenge on Young's structural injustice argument.

[37] Barry and MacDonald make a similarly powerful argument against Young, in that 'what matters is not that the harm has been intended, but merely that it could reasonably have been foreseen, and that the agent could have avoided acting in the harmful way without incurring great cost' (2016: 99).

structural injustice exists outside the liability model. As I shall try to demonstrate later, such a distinction is key to addressing structural injustice.

First, the harmless torturer example does not capture the processes that cause structural injustice on Young's account. Someone in the harmless torturer story is responsible for rigging up voltage charges through an incarcerated man strapped to a torture machine. This is far from the ordinary conditions of structural injustice that Young was trying to describe. As she says, the social connection model of political responsibility 'does not evaluate harm that deviates from the normal and the acceptable; rather it often brings into question precisely the background conditions that ascriptions of blame or fault assume as normal' (2006: 120).[38]

Second, I suggest that the connection between the tortured man and the agents who pushed the switches is too linear to be a structural process in the way that I think Young intended. In Sangiovanni's example, the switch-pushers' connection of action to injustice is clearly traceable and therefore captured by the liability model under Young's account of strict liability. Those individuals who flicked the switch are directly connected to the act of torture, albeit initially ignorant of this fact. Even though each individual may not have caused enough charge on their own to inflict torture, the premise of this example is that we already know that collectively this particular group of people is directly traceable. These connections not only imply straightforward liability on the part of whoever had wired the victim up to the switch-pushers but also that the prevention of further harm can be accomplished by all concerned with some fairly commonsense alterations to their lighting arrangements. The amorphous nature of the structural processes that enable structural injustice is not captured by the harmless torturer example. Here the key point is that we should understand our connections to structural injustices as being imbricated in the mere participation in daily life. As McKeown explains, 'An individual may not contribute in any significant way to the background structure, but simply by acting within it the individual is reproducing those structures' (2018: 500). In this sense, Sangiovanni's 'reasonable alternative' condition, whereby individuals could simply choose a different action, is not met. Individuals alone cannot create alternatives to the ordinary and

---

[38] Also see Young (2011: 107).

legitimate practices of everyday life. Change to structural injustice must come at the macro level.

At first then, Sangiovanni's claim seems intuitively right that once a particular social group (e.g., a racialised or gendered group) is disadvantaged by the actions of others (no matter how diffuse or unintentional), this should be understood as indirect discrimination, whereby seemingly legitimate behaviour has adverse effects on a particular social group.[39] However, because forms of connection to a social process serving structural injustice are untraceable to particular individuals, groups or institutions, they remain distinctly structural rather than identifiable acts of discrimination (albeit unintentional). Whether direct or indirect, discrimination 'is primarily an agent-oriented, fault-oriented concept' (Young, 1990: 196). As Young explains:

> [The concept of discrimination] tends to focus attention on the perpetrator and a particular action or policy, rather than on the victims and their situation ... In its focus on individual agents, the concept of discrimination obscures and even tends to deny structural and institutional frameworks of oppression. (1990: 196–197)

A different sort of critique is offered by Nussbaum (2009, 2011). Her interpretation of Young's concept of responsibility is grounded in the following assumptions: 'An agent is responsible ... only if a) the agent is causally embedded in processes that produce a problematic result and b) the agent is in a position to assume ongoing forward-looking responsibility (in cooperation with others) for ameliorating those conditions' (Nussbaum, 2011: xx). I argue that first, in (a), the term 'causally embedded' is understood by Nussbaum not as Young intended, but rather as a linear link to injustice (much like Sangiovanni). Second, Young does not require (b) to establish responsibility for structural injustice. Indeed, Young is explicit on this point: 'I [have] resisted the suggestion that different people bear different degrees and kinds of responsibility as contributors to structural injustice. To say that responsibility is shared means that we all bear it personally in a form that we should not try to divide and measure'

---

[39] Social groups are often represented in law as those with 'protected characteristics' (see, e.g., the UK Equality Act 2010).

(2011: 124). I suggest that these two features of Nussbaum's critique lead her (and those who echo her perspective, such as Goodhart (2017)) to mistakenly claim that Young's account cannot escape retrospective liability for structural injustice:

> It seems to me that what we ought to say is that if person A has a responsibility . . . [for structural injustice], and she fails to take it up, then, when the relevant time passes, she is guilty of not having shouldered her responsibility. I think that this follows quite simply from the logic of ought. Young says that A ought to shoulder the burden; well, that appears to imply that if A does not shoulder the burden A has done something wrong. (Nussbaum, 2011: xxi)

I suggest this is to misunderstand the political dimension of Young's argument. Thinking back to Young's story of Sandy, consider the housing market in an affluent city where competition for scarce housing drives prices up, and many individuals and families are priced out. Now consider the story of Rosa who has been working for a large multinational company as a technician for fifteen years and is based on the outskirts of Reading. Rosa receives notification that her company is axing 60,000 jobs over the next two years due to the decision to phase in AI to substitute large numbers of human occupations in the organisation. Rosa's job is one of those to be axed. She is extremely concerned about her financial future as well as providing for her children. Rosa begins looking for another job and after several months manages to get one working for a small firm in the centre of London. Rosa and her family decide to relocate and she rents a house to live in with her children in the suburbs of London. Rosa did not intend to contribute to the city's inequality but was constrained by the fact that she needed to live in commuting distance of her new job in central London. Has Rosa done something wrong in renting a house in a city where some people are very poor and in the worst cases homeless? What ought she to do? She could campaign for affordable housing, rent-setting and more state-provided accommodation. She might persuade others to join her to effect change. This sort of collective civil action could have a powerful impact and is certainly worth doing. But is Rosa *at fault* if she does not do these things or perform acts like them, beyond voting in elections for councillors and other politicians who pledge to address homelessness on her behalf, and with greater resources and coordinating power than Rosa has at her disposal as an individual? The structural matrix here is a complex mix of the relative social positioning of millions of people to each other as well as labour patterns, life cycles, governance and legal frameworks, ownership, and (legitimate) pursuits of interest – to name just a few of the many dynamic structural variables involved. Tracing a causal link between Rosa's decision to rent a house in London and homelessness would prove impossible on the individual level: it is not at all clear how her city residency contributes to the complex circumstances of particular cases of homelessness (which are perhaps influenced by familial poverty, failures of educational institutions, an unstable or exploitative economy, abusive personal relationships, mental health issues, etc.). But even if you did think Rosa was at fault, in the sense that her rent-paying is part of a macro-level set of unequal social relations, what about all the other forms of structural injustice beyond homelessness in London? How many is it reasonable to ask Rosa to actively take up at the same time – 2, 5, 10,000, 20,000? On Nussbaum's account, the answer to this question would have to be – *all* of them.

To demand this scale of attempted tracing would be to render the exercise of politics impossible. Here we can think back to Ali's interview with Webb in the Introduction: 'I'm not complacent – I'm exhausted, I'm overwhelmed! . . . I do not really understand this stuff. Even if "I see something and I say something", no one's going to care.'

Nussbaum's implausible requirement – which I will call the 'liability scale problem' – is inherent to the liability perspective in the context of structural injustice. It goes to the heart of why defining structural injustice as distinct from the sort of blameworthy injustices captured by the liability model is so politically important. Rosa is responsible in the sense that she has contributed to the production and reproduction of the background conditions of a particular structural injustice, she may even be thoughtless but she is not liable. This distinction is vital for thinking through different political approaches to different forms of injustice.

IMPERFECT POLITICAL RESPONSIBILITY

Nussbaum and in fact most of Young's critics argue that, in some way or other, structural injustice is the consequence of a failed duty. In Young's account of structural injustice there are contradictory references to the notion of duty,[40] but the following explanation is the most useful in developing a picture of her perspective on political responsibility for structural injustice:

> Like duties, responsibilities carry a burden and an obligation. Unlike duties, however, responsibility carries considerable discretion. One must carry out one's responsibilities, but how one does so, is a matter for judgement according to what the responsibilities are for, the capabilities of the agents and the content of action. (2011: 143)

That is to say, Young's concept of political responsibility is *discretional*: '[i]t is up to the agents who have a responsibility to decide what to do to discharge it within the limits of other moral considerations' (2011: 143). Rather than a responsibility of duty then, Young's conceptualisation of political responsibility for structural injustice is a relational call for solidarity that is aimed at those who contribute to the background conditions of structural injustice, often including those who are struggling against structural injustices themselves and are connected through those injustices to many others (McKeown, 2018).

What is more, as Jugov and Ypi argue, is that because all those who contribute towards the background conditions of structural injustice include those most negatively affected, it is important not to invoke 'very stringent duties that demand to be discharged at a very high cost to the victims of structural injustice' (2019: 22). The idea that there is a conceptual difference between political responsibility for structural injustice and duty immediately begs the question – what kind of responsibility is political responsibility, if not a duty? Somewhat ironically, I find an answer in the work of one of Young's critics.

Reiman (2012) provides a critique of Young, which, while on the whole unpersuasive to me, nevertheless raises a very helpful way of answering the question of what sort of responsibility is at the centre of Young's account. Reiman argues that because Young did not develop a theory of justice, as such, she is unable to discern an injustice from bad

---

[40] See Young (2011) for example, where on page 76 she says that political responsibility is a duty, and yet on page 143 she explicitly argues that it is not.

luck and that approach could be no more than a charitable response to misfortune. Reflecting on Young's figure of 'Sandy', Reiman suggests: 'We might think of her not as a victim of injustice, but as a person who has suffered a misfortune for which she is not to blame. And then moral responsibility might be evoked on the basis of something like a duty to lend innocent sufferers a hand' (2012: 743). This part of Reiman's argument misses the importance of 'action' in Young's account. What motivates responsibility for structural injustice is one's contribution to the structural processes that serve as the often predictable and routinised background conditions of structural injustice, and not merely pity for another's random misfortune.[41]

Reiman's thoughts on charity lead him to the idea that we ought to think of responsibility for structural injustice as a sort of 'imperfect duty' whereby individuals *do whatever they can* rather than having to act according to a defined set of duties as Nussbaum, Goodhart, Sangiovanni and others hold. The idea of addressing structural injustice as an imperfect duty in Reiman's work is only mentioned in passing, but has great potential.

As Young intended, one's responsibility will depend on one's place in the social connection matrix and correlatively on one's resources to effect

---

[41] This sort of criticism has some similarities to that of Ronald Dworkin's resourcist theory of justice. Dworkin was concerned with the redistribution of resources according to one's luck, of which there are two forms. The first is 'option luck', resulting from the consequences of one's efforts, talents and judgement; the second is 'brute luck', which falls on individuals irrespective of their behaviour. Dworkin explains: '[i]f I buy a stock on the exchange that rises, then my option luck is good. If I am hit by a falling meteorite whose course could not have been predicted, then my bad luck is brute' (2002: 73). On this view, bad brute luck deserves compensation and bad option luck does not. Such a liability approach leaves no room for a structural perspective on injustice, in which responsibility is taken up for 'the social processes that open or limit opportunities for individuals to flourish' (Young, 2011: 39). Indeed, for Dworkin, because a great deal of inequality in the present is justified by past judgements and actions, it therefore falls outside the realm of political responsibility altogether. As Wolff (2018) has argued, political theory has been dominated by a particular sort of justice paradigm, emanating from perspectives such as Dworkin's, which fail to capture a great many injustices – structural injustices. Young, he suggests, offers the beginnings of an alternative approach and indeed that is the premise of this book. In a similar vein, Haslanger (2024) argues that 'The mainstream philosophical literature is of limited help, for it is mostly committed to an outmoded methodological individualism.'

change. In this way, and having developed a structural perspective, sometimes this will amount to little or nothing among the huge array of structural dynamics at play in any given context, but sometimes it will result in concerted action on a particular issue. While resisting the idea that we should measure the extent of each person's degree of political responsibility (Young, 2011: 124) and bearing in mind Young's rejection of 'duty' in the context of structural injustice, the idea of 'imperfect responsibility' facilitates collective political action and overcomes the liability scale problem inherent in Nussbaum's account of structural injustice.

The idea that responsibility for addressing structural injustice is an imperfect responsibility chimes with a similar argument regarding imperfect obligations to meet corresponding human rights which I find illustrative here. Sen (2004) argues that while it is impossible for individuals to carry out all that would be required to meet everyone's human rights (as set out in the Universal Declaration of Human Rights, for example), what is needed instead is for each of us to assess priorities and to 'give reasonable consideration ... followed up by sensible choices of action' (339). This is not, he argues, 'an agreement to tie oneself up in hopeless knots' but instead 'each person has to judge the extent to which they can make a difference' and while '[a] great many parametric considerations ... will inescapably figure in the reasoned evaluation of what a person should do', this approach is much more productive than 'proceeding on the assumption that we owe nothing to others, unless we have actually harmed them' (2004: 339, 340).[42] In recent work, Zheng similarly suggests that we adopt 'moral aspiration ... to regularly take up at least some opportunities to help others, contribute to social change and so on' (2021: 508).

In this sense, imperfect responsibility is an effective description of what I think Young understood to be political responsibility. My view is that it is sufficient to generate discretionary action for addressing structural injustices without requiring the stricter moral or legal requirements of the liability model and hence the liability scale problem raised by Nussbaum's critique of Young.

[42] See Browne (2013b) for detailed discussion; also Sen (2004).

IF NOT LIABILITY, THEN BENEFIT?

Rather than focusing on liability, some have argued that we should instead trace who benefits from structural injustice in order to ground responsibility for addressing it. For example, Wolff (2024) explores what he sees as the difference between structural injustice and structural harm with the distinction resting on who benefits. However, as I shall explain, for reasons that mirror my critique of those who argue for addressing structural injustice with liability-based solutions, I am not sure that the tracing of benefit is necessarily productive.

Wolff begins with the uncontroversial argument that while all injustices are harms, not all harms are injustices. Certainly, this makes sense when thinking about legal or moral liability. Thinking back to the case of the 'self-driving car' in the Introduction, one may suffer an injury or worse still, die, as a consequence of someone else's negligence. While unintentional, we might nevertheless attribute blame for a harm rather than claim injustice. However, this logic, I suggest, does not extend to the structural context. Wolff argues that if an individual, group or institution is structurally harmed *to the benefit* of another individual, group or institution, this should be understood as a structural injustice, and if no benefit is gained then this should be understood as a structural harm.

Certainly, there is considerable ambiguity in Young's work, and it could be seen to imply Wolff's interpretation when Young says: '[s]-tructural injustice, then, exists when social processes put large groups of persons under systematic threat of domination or deprivation of the means to develop and exercise their capacities, at the same time that these processes enable others to dominate or to have a wide range of opportunities for developing and exercising capacities available to them' (2011: 52). Indeed, Young defines 'structural social groups' as the relative positioning along axes of disadvantage and privilege within the structural matrix.[43] Maboloc expresses a similar view when he says that '[s]tructural imbalance causes injustice' (2019: 1189). My interpretation of Young's words, however, is that she is describing the relative social positionings of different individuals, groups and institutions within the

---

[43] See Young (2009).

structural matrix, rather than a strict definitional distinction between structural injustice and structural harm. It strikes me that for the same reasons that Young's social connection account of responsibility does not entail the tracing of blame, structural injustice does not entail the tracing of benefit. Such a requirement would serve to ground political response in the activity of tracing benefits rather than focusing on addressing the structural processes that cause structural injustice. What is more, when we focus on ascertaining the benefits of individuals and institutions as the definitional point of structural injustice as Wolff suggests, we are also drawn to the consideration of corresponding redistribution policies by way of remedy. I think that Young's point (and one I have sharpened by my selective descriptions of Young's work) was to operate on a different register to the exercise of tracing and redistribution in the specific context of structural injustice:

> Rectifying injustice ... implies not compensating people for disadvantage or engaging in policies of redistribution after the social processes have wrought their damage. Instead, promoting justice in social structures and their consequences implies restructuring institutions and relationships to prevent these threats to people's basic well-being. (Young, 2011: 34)

While of course it is important to have a sense of whose social structural position might benefit from the demise of another in order to think through what is needed by way of redress, the central feature of structural injustice is that people are made vulnerable to domination, exploitation and exclusion as a consequence of structural actions and through no wrongdoing. What defines structural injustice is one's positioning in the structural matrix and one's corresponding life prospects, not a calculation of others' benefits per se. So, if we do not distinguish between negative outcomes that seem to benefit no one on the one hand and negative outcomes that do unwittingly benefit some on the other, where does that leave our understanding of structural harm? I argue that structural harm is no different from structural injustice. Rather than diverting our political attentions to the tracing of benefits to some when what is needed is collective action to change the background conditions of structural injustice for all those affected (Young, 2011: 45). I therefore use the terms interchangeably as does Young: 'Structural injustices are harms that come to people as a result of structural processes in which many people participate' (2003: 7).

POLITICAL TRANSITION AND RESPONSIBILITY

As we have seen, Young developed her seminal account of political responsibility for addressing structural injustice as an additional and distinct form of responsibility to that which is needed for liability-based injustice, through a critical engagement with the work of Arendt:

> To the extent that we participate in the ongoing operations of a society in which injustice occurs, we ought to be held responsible. This does not, however, make us guilty or blameworthy or directly liable for paying compensation to victims of harm. Responsibility in that liability sense should be reserved for persons who can be specifically identified as causing the harm ... I think Arendt is right that it is a bad idea to blur the conditions for saying persons are guilty or at fault. It is often important to pin a wrong on someone who did something or was in the unique position to have prevented it. This means that we should conceptualize responsibility differently for the purposes of assigning responsibility for structural injustice. (Young, 2011: 104)

I have argued that the very point of Young's account of political responsibility for structural injustice is *not* to let everyone who is not traceably guilty 'off the hook'. Instead, it is to ground those who contribute to the background conditions of structural injustice in a responsibility to address those conditions. This is to require political redress for that which tends to be understood as either individual misfortune or the consequences of individual irresponsibility rather than of those macro-level accumulations of actions that I have called structural actions. This sort of approach demands more of a forward-looking task, although it is sometimes necessary and informative to look backwards in an attempt to see just how these structural processes might have come into being (Young, 2011: 108–109). Lu (2017), for example, makes a convincing argument that historical injustices such as colonial practices ought not to be redressed simply through notions of accumulated moral debt handed down to contemporary agents, but that the continuity of reproduced unjust social structures rooted in the past ought to be a collective target for contemporary political responsibility. Similarly, I think Zheng is right to argue that '[t]he distinctive difficulty raised by structural injustice is that injustice is not an "isolatable action or event that has reached a terminus", but rather "is ongoing, and is likely to persist unless social processes change"' (2019: 111).

Key to my defence of Young's argument that structural injustice is not captured by a liability-based approach is the sharpening of the idea that structural actions are untraceable. However, even though I see political responsibility to address structural injustice as an imperfect responsibility rather than a duty, I disagree with Young that political responsibility ought to be understood as wholly distinct from liability-based (legal and moral) responsibility as described earlier. I do not do so, though, for the reasons given by Nussbaum, Sangiovanni, Barry and MacDonald or other advocates of liability-based approaches. My sense is that Young's account rendered a position on political responsibility she perhaps did not intend but nevertheless produced. As I have discussed, having distinguished liability-based injustice from structural injustice, Young assumed that the form of responsibility for each must also be distinct. Her formulation of what sort of responsibility is required to address structural injustice is set in contrast to that which is required for the liability model: 'this distinction is a matter not of degree, but of kind' (2011: 92). It seems implausible to me, however, to restrict such a seemingly large concept – 'political responsibility' – to the sole remit of structural injustice. This is a very different sort of critique to that usually aimed at Young's work. Young's use of the word 'political' in 'political responsibility' was meant to suggest the coming together of peoples to act against structural injustice – the responsibility to be collectively political – and that such a feat could not be enacted by an individual, such as in the example of an individual abiding by a law or moral norm in a given context. My suggestion, however, is that political responsibility is surely applicable to addressing both liability-based and structural forms of injustice as a sort of precursor to change. For example, listening to a speech to state leaders by the Swedish activist, Greta Thunberg, at a UN Climate Change Conference, it is hardly surprising she inspired children and adults the world over to join her environmental campaign:

> You only talk about moving forward with the same bad ideas that got us into this mess, even when the only sensible thing to do is pull the emergency brake … We have come here to let you know that change is coming whether you like it or not. The real power belongs to the people. (2018)

Thunberg is publicly exhorting people to join her to effect change, to stop converting natural resources into commodities and rather to appreciate them as a balanced ecosystem of which we are part. In doing so, she asks us to take up the responsibility to be political in Young's sense of political responsibility. However, there are two important observations to make about the practice of political responsibility in this example. The first is that while Thunberg epitomises Young's idea of adopting a structural perspective in order to enact the sort of political responsibility discussed earlier, Thunberg's call to action is not restricted to structural injustice – the untraceable unintended consequences of the actions of millions of people, groups and institutions who, in virtue of their structural actions, contribute in a multitude of ways to the environment's degradation. Thunberg's campaign is also a call to individuals, groups, corporations and other institutions as well as states to stop committing what we already can determine as direct and indirect wrongful abuses of the environment, and, if possible, to 'pull the emergency brake' as a moral act of responsibility if not a legal one. Thunberg is enacting political responsibility in calling others to abide by laws and moral norms or to create new ones in favour of the environment's health. The main point here is that Thunberg's objective is in fact to establish liability (particularly including those leaders with the most power to effect change) but her method is the enactment of political responsibility. Recall that in Young's account, liability is addressed through legal and moral responsibility, not by what she calls political responsibility (a 'distinction is a matter not of degree, but of kind').

By broadening the definition of political responsibility whilst also retaining the distinction between liability-based injustice and structural injustice, we can address this conceptual problem. In the case of Rosa, for example, although she is not liable for the homelessness that surrounds her, a multitude of social changes could become part of a long-term solution (these could be anything from free public transport during certain hours to resources that enable activities to create a stronger community spirit). These would not produce a particular fix for homelessness but to develop an awareness of the background conditions of the structural elements of a problem like homelessness is likely to have a positive impact nevertheless. This is an argument for developing a structural sensitivity rather than limiting oneself to political solutions grounded in traceable liability (albeit recognising their vital importance). Indeed, I suggest we could rename what Young called 'political responsibility' as discretionary 'structural responsibility', solely focused on the exercise of speculating and addressing the background conditions of structural injustice, and consider it a subset of the wider concept of political responsibility that pertains to both liability-based and structural injustice.[44] Because structural injustice is untraceable, structural responsibility is best focused on the background conditions of structural injustices.[45] As I shall discuss in Chapter 4, this requires the engagement of Arendt's 'everybody' in political fora designed to bring structural speculation directly into governance mechanisms and a reweighting of interests in policy design and regulation.

Relatedly, my final observation on Young's account of structural injustice, is that it is rather static in its characterisation of different forms of injustice. That is to say, that there is no obvious articulation of transition between structural and liability-based injustices in Young's work. I want to make the argument that although the different forms of injustice are distinct, there is nevertheless a possibility of transition between them, and that this is politically important. Unlike Sangiovanni's argument, which highlights ignorance (specifically indirect discrimination) as the cause of our inability to join the dots of our actions to injustice, I want to suggest something quite different – a

---

[44] Beck (2020) developed a concept of structural responsibility also. However, his interpretation is different in that Beck positions 'structural responsibility' as an alternative to 'political responsibility'.

[45] In Appendix 1, I have charted the various terms used in this conceptual chapter and illustrated how they relate to one another in Tables 1 and 2.

transition from what was previously untraceable to the traceable. For example, David Attenborough (2019: np) explains that there was a time historically when 'we did not know that we were changing the climate'. The difference is that we *could not* have known that we were changing the climate – our structural actions were untraceable. Only with more recent methodical scientific research have we come to understand humanity's harmful impact on the planet after much scientific speculation (UKRI, 2022). As a consequence, our methods of speculation and exploration improve the possibility of tracing our individual structural actions to injustice such that they may become identifiably linked. Once these traces have been established, our structural actions transition into the realm of traceable moral responsibility (actions that abide by moral expectations and norms pertaining, for example, to the environment, such as recycling plastics). From here, through forms of political deliberation and other forms of political action such as campaigning and protest, we may transition those moral responsibilities into traceable legal requirements such as banning the use of plastics in certain contexts altogether (legal responsibility). In this way a transition occurs whereby structural actions become morally unacceptable and perhaps legally unacceptable actions. Another example is that of the vast data analysis of machine-learning techniques that hitherto were impossible to comprehend but now show new patterns and consequences of our behaviour, whether it be through face-recognition surveillance or wearable technology measuring the activity of hundreds of millions of people at a time. More of that further on but suffice it to say here that our structural perspectives on the consequences of our connections to structural injustice may, in time, be met with new forms of knowledge that render a transition of structural actions into moral or legal actions which are captured by the realm of liability.

Political solutions to structural problems are currently lacking, and the first step towards improvement is a subtler appreciation of their shape and distinctiveness. However, and of crucial importance politically, even with great leaps in technological advancements, there may be no epistemological possibility of transitioning to a liability-based solution for some structural injustices. That is to say, that we may never have a clearer picture than a dim perception that our actions are in fact structural actions contributing to the background conditions of a particular structural injustice, or one of many, in an inconceivably convoluted macro scale. We ought nevertheless to attempt to address both the background conditions of structural injustices as well as negative structural outcomes without restricting our energies to tracing fault; thinking back to Rosa's story again, a given society reflecting on its structural actions, might provide free accommodation for those in the most need with no course to traceability – period.

My intention in this first chapter has not only been to defend the idea that structural injustice is distinct from fault-based forms of injustice but also to sharpen the distinction found in Young's account. Such a distinction is intended as a foundation for later chapters where I attempt to find practical political tools to address contemporary structural dynamics in the context of AI and Repro-tech, both technologies set to greatly impact future generations. Young sought to capture a certain sort of injustice that existed outside established legal and moral categories of liability, providing an explicit frame to enable and focus political redress. As I have tried to demonstrate, many scholars have rejected the distinction between structural injustice and fault-based injustice, instead attempting to subsume structural injustice within the liability model of political responsibility. However, ultimately this insistence only would serve to return us to the problem that Arendt posed and that, I have suggested, was so inspirational to Young – 'where all are guilty no one is to blame'. Our ability to change instances of structural injustice is not best served by turning our political energies towards tracing fault backwards – not until, at least, we have the epistemological tools necessary to make such a task meaningful. This is not to exculpate those who have an untraceable causal relationship to injustice but rather to do the opposite – to bring the structural dimension of 'ordinary injustices' *within* the remit of political response irrespective of their causal complexity. Determining what we are responsible for (rather than liable for) ought to involve an exercise in speculating on the social structures in which we are embedded and the ways in which our actions both affect and are affected by those structures. Structural actions are imbricated in the places we live, the jobs that we do and travel to, the institutions we rely on and the technologies we develop and use. Liability ought not to set the limits of our political ambitions to address the everyday injustices that emanate from these actions.

In the following chapters, I turn to the context of transformative technologies, AI and Repro-tech, to explore how the ideas set out in this discussion might operate.

CHAPTER 2

Artificial Intelligence and Ground Truth

Artificial intelligence is the future . . . for all humankind . . . Whoever becomes the leader in this sphere will become the ruler of the world.

Putin (2017)

This huge technological revolution is in fact a political revolution.

Macron (2018)

I N THIS CHAPTER, I WILL EXPLORE WHAT POLITICAL responsibility for the governance of AI currently looks like. Certainly, there is a macro-level keenness to facilitate the predicted leaps in GDP[1] that new AI innovations promise to bring. At least sixty states have placed AI at the centre of both domestic and foreign strategies.[2] Set against a backdrop of remarkable technological innovation,[3] however, is a deep public anxiety about the prospect of AI-generated harm and AI substitution of human labour. States, international organisations and corporations have attempted to assuage these fears with a raft of

---

[1] Recall AI is estimated to contribute $15.7 trillion to the global economy by 2030 (World Economic Forum, 2022).

[2] According to OECD (2021), sixty countries have recently developed some form of AI strategy: Argentina, Australia, Austria, Belgium, Brazil, Bulgaria, Canada, Chile, China, Columbia, Costa Rica, Croatia, Cyprus, Czech Republic, Denmark, Egypt, Estonia, Finland, France, Germany, Greece, Hungary, Iceland, India, Ireland, Israel, Italy, Japan, Kazakhstan, Korea, Latvia, Lithuania, Luxembourg, Malta, Mexico, Morocco, Netherlands, New Zealand, Norway, Poland, Portugal, Romania, Russia, Saudi Arabia, Serbia, Singapore, Slovak Republic, Slovenia, South Africa, Spain, Sweden, Switzerland, Thailand, Turkey, United Arab Emirates, United Kingdom, United States, Uruguay, Vietnam, as well as the European Union. For some specific examples, see, US Federal Government (2023); HM Government (2021a); UNESCO (2021); EUR-Lex (2021).

[3] See, e.g., Taddeo and Floridi (2018).

guarantees that 'Responsible AI', human oversight, transparent AI and a surge in new AI sector jobs will counter any downsides to the new 'AI-age'. In this chapter I shall consider these themes in relation to the conceptualisation of structural injustice discussed in Chapter 1. I suggest that it is politically imperative that the structural dynamics of AI are taken seriously, and demonstrate how current governance approaches to AI, which are principally grounded in liability, are woefully insufficient.

AI'S TRUTH

[W]e can safely dispense with the question of whether AI will have an impact; the pertinent questions now are by whom, how, where, and when.

Cowls and Floridi (2018: 1)

A great deal has been made of the fact that today many decisions are allocated to algorithms without direct human involvement: performing medical diagnoses, determining whether or not a person is creditworthy, predicting who is the right candidate for a particular job or calculating how long someone ought to be sentenced to prison.[4] Framed by the global context of targeted economic growth and attendant strategies of efficiency, the human is a poor substitute for the vast computational power of machine learning, the creative capabilities of generative AI and the speed, precision and regularity of automation in so many spheres of human activity. We only need to think of the Google search engine by way of example: if I pose the question 'Where is the planet Mars?', 166,000,000 page-ranked results are generated in 0.88 seconds. It goes without saying, no human could attempt such a feat. Similarly, ChatGPT has thrilled users with its generative ability to create original material in any genre, in any language, in a matter of a few seconds. When I asked for a poem on 'The governance of AI in the style of Shakespeare', ChatGPT produced the following line describing AI: 'For thou art a child, born into the world anew, With no knowledge of right or wrong

---

[4] As the seminal works of Haraway (1985), Hayles (1999) and Braidotti (2013) led us to imagine, humans have become co-constituted by technology. Depending on your view, we have become more than, or less than, human.

or what to pursue, And so it falls upon us.'[5] More impressive still has been the creation of AI-generated original images, music and voice-mimicking which are of extraordinary quality and increasingly impossible to distinguish from human creativity.[6]

AI is a term that encompasses a range of technologies that use algorithms to perform tasks or solve problems that previously only human intelligence could attempt. Machine learning, as a subset of AI, is at the centre of what makes a system 'intelligent', in that these algorithms learn from structured labelled data to make predictions and continually adjust their interpretations accordingly.[7] Selected inputs are weighted according to relevance or importance to determine an output – an algorithmic decision.[8] At the baseline of this process, human judgement is crucial and as Amoore's (2019) work shows, we can think of

---

[5] While far from the quality of Shakespeare's writing, it is an astonishing attempt nevertheless and completed in less than 10 seconds: 'Oh wondrous marvel, born of man's design, Thy circuits and thy wires doth gleam and shine, But with great power comes great responsibility, And governance of thee requires great ability./ For though art like a child, born into the world anew, With no knowledge of right or wrong, or what to pursue. And so it falls upon us, the masters of they fate, To guide thee on thy path, lest thou should meet a dire state./ With laws and regulations we must rein thee in, And teach thee morals, lest thou should cause great sin, And though thou may protest, and claim to be free, We know thy limits, and shall govern thee wisely.' ChatGPT (10 April 2023).

[6] See here, e.g., a project by Ing and Microsoft entitled 'The next Rembrandt' which posed the question, 'Can the great master be brought back to life to create a new painting?' and then answered it with the production of a remarkable AI-generated original 'self-portrait' in the style of Rembrandt: www.youtube.com/watch?v=IuygOYZ1Ngo

A darker story is that of Jennifer DeStefano from Arizona who received a call seemingly from her distressed daughter, Briana, claiming she had been kidnapped, followed by a demand for money. It turned out that the voice of Briana had been cloned from social media by scammers using AI. The kidnapping was a deepfake but wholly convincing at the time of the call (Salam, 2023).

[7] Despite this definitional point, I tend to use AI and 'machine learning' interchangeably in line with common parlance.

[8] Watson and Floridi (2020: 9216) is right to point out that there is no real sense in which machines take 'decisions': This, he says, is an 'anthropomorphic trope granting statistical models a degree of autonomy that dangerously downplays the true role of human agency in sociotechnical systems'. Because machines cannot 'think' in the way that humans do (Floridi et al., 2009; Floridi, 2017), we are ultimately responsible for the outcomes generated by AI even when we cannot understand the processes by which AI provides the data we ask of it. That said, for reasons of clarity I will use the phrase 'algorithmic decisions' in this article with Watson and Floridi's point in mind.

Amazon's Mechanical Turk, as an example of the foundations of AI.[9] This is a commercial crowdsourcing service that provides companies with a vast virtual human workforce for the construction of training data for machine learning in the form of manual data validation, content moderation and the creation of a hierarchy of different data inputs. That is to say, a human 'data tagger' has determined the characteristics that distinguish and label different data inputs and their weighted relative value – the 'ground truth' – for a given objective (Muller et al., 2021).[10] Due to the human intervention involved in the construction of a ground truth dataset, this sort of machine learning is classified as 'supervised learning'. However, deep neural network machine learning – 'unsupervised learning' – is a step further towards autonomy. This sort of AI is able to move beyond the structured and labelled training data to distinguish vast amounts of unstructured and unlabelled data by observing and reinterpreting data patterns at astonishing speed with no human intervention. Ultimately, this 'deep learning' AI can create its own ground truth for a given objective, albeit with some inherited implicit values. Amoore (2019: 151) argues:

> The claim to truth made by machine learning algorithms, then, is not one that can be opposed to error or falsity ... The architectures of neural network algorithms can contain multiple hidden layers, hundreds of millions of weights, and billions of potential connections between neurons ... Put simply, the mode of truth of the algorithm is entirely contingent on a particular notion of ground truth in the data.

---

[9] See the Mechanical Turk website, www.mturk.com/. Human zero-hours contractors conduct what Jeff Bezos (CEO of Amazon) calls 'artificial artificial intelligence' (Jones, 2022).

[10] However, and somewhat ironically, some human data taggers have started to utilise generative AI to increase their income. Dzieza (2023) discusses the low-paying jobs of data-taggers and the story of one Kenyan data tagger who decided to innovate: 'after his account got suspended for mysterious reasons, he decided to stop playing by the rules. Now, he runs multiple accounts in multiple countries, tasking wherever the pay is best. He works fast and gets high marks for quality, he said, thanks to ChatGPT. The bot is wonderful, he said, letting him speed through $10 tasks in a matter of minutes. When we spoke, he was having it rate another chatbot's responses according to seven different criteria, one AI training the other.'

As Cassie Kozyrkov, Chief Decision Scientist at Google, describes ground truth (2020: np), 'there is no "*right*" answer. The "*right*" answer depends on what the owner of the system wants the system to do ... In AI the objective is always subjective.'

Here I find Kozyrkov's example of an algorithmic system tasked with identifying a 'cat' illuminating. Simulating the system with a human audience, Kozyrkov presents a collection of images and asks the audience to identify each as either 'cat' or 'not cat'. Accordingly, the images of cats are identified as 'cat' and the images of dogs, hamsters and the like are identified as 'not cat'. Finally, Kozyrkov asks the audience to identify the last image – a tiger. Here the audience are unsure and split in their opinion. '"*Big cat*"? "*Sort-of-cat*"? "*Maybe cat*"? Those aren't allowable options! You're a system that is programmed to only output "*cat*" or "*not-cat*". So which one is it?' asks Kozyrkov. Unable to agree, the audience is then informed that the algorithmic system they have been tasked with simulating is a 'pet recommender system which only suggests creatures which are safe to cuddle in their typical adult form'. Now the answer becomes clear – 'not cat'.

The importance of Kozyrkov's example is not only to expose the subjectivity built into seemingly objective AI systems but also to highlight the impact of AI 'inheritance'. That is to say, the ground truth of one or more algorithmic systems produces data that is subsumed into the training data for another and another, all repurposed from their original task. It turns out, as Kozyrkov explains, 'Ground truth is not *true*' (2020: np).

AI-GENERATED HARM: LIABILITY BASED AND STRUCTURAL

Artificial intelligence as an exacerbator of human bias is fast becoming a common story. Algorithms tend to reduce the complexities of human society to small numbers of data points on a scale incomparably larger than ever seen before. We can think about the implications of this sort of crude simplification through the more abstract work of scholars such as Amoore and Butler, who remind us that there is always an inherent partiality to any perspective, even of oneself. As Butler explains; 'my account of myself is partial, haunted by that for which I have no definitive story ... a certain opacity persists' (2003: 29). AI, however, gives the impression that the data it has selected as a particular output has been definitively identified – that is to say, it has been somehow reduced down to an essence and 'placed beyond doubt' (Amoore, 2019: 151). We ought to remain conscious that algorithms do not and indeed, could never, serve to bring ourselves and each other into sharp focus. Like humans, AI cannot present the objective truth about a person's perspectives, beliefs and behaviours.[11] In this sense algorithms 'illuminate the already present problem of locating a clear-sighted account in a knowable human subject' (Amoore, 2019: 150). The reductive quality of AI serves to weaken our ability to recognise the heterogeneity of behaviours, thoughts and possibilities.[12] Butler's work highlights how we tend to seek comfort in what we think of as surety in the form of basic categories and simple labels, and yet in doing so we damage something of humanity by not challenging the discourses that have become our 'epistemological and ontological anchor' (2004: 35). This connects to Arendt's idea of thoughtlessness, discussed in Chapter 1, whereby in our need to submit to habitual behaviours that help us function in the everyday world, we nevertheless leave the hard labour of scrutiny and critique of our thoughts, beliefs, actions and habits to one side. In amongst the many ways AI can help humans to discover new patterns of behaviour and other phenomena, it also has the capacity to greatly exacerbate thoughtlessness.

A simple extension of this point is the stark example of AI recruitment algorithms. This is a technology which has already transformed the human resources sector in many countries (Drage and Mackereth, 2022). Unsurprisingly, high-profile companies are often faced with very large numbers of applications for jobs and are increasingly attracted to what is known as 'intelligent screening software'. This sort of AI uses companies' existing employee data as its training ground and calculates

---

[11] This is irrespective of how many methods of metrology (the science of measurement) are applied to the quantification of 'uncertainty' in AI outputs (Levene, 2022).

[12] It is also vitally important that, as Braidotti argues (2022), a progressive human politics must, to a much greater degree, de-centre the bios (the 'exclusively human life') so as to better engage with the zoe (non-human life) and geo (planetary and environmental) as well as recognise the place that techno (science-derived non-living) forces will increasingly play in the future.

which existing employees have become 'successful' within the company and which have not; that is to say, the ground truth of HR needs is calculated. Such ranking is built upon metrics of education, skills, experience and, most importantly, key performance indicators set by the company. From those findings, the algorithm can map certain characteristics to CV data submitted by new applicants as well as extend an applicant's data with a trawl of their social media profiles and any other publicly available personal data. From there the algorithm can weed out those who are unsuitably qualified or unlikely to become as successful as the 'ideal employee', in line with the profiles of current employees (Ruby-Merlin and Jayam, 2018; Saundarya et al., 2018). Consider the extraordinary example of leading technology firm Amazon, which abandoned its home-grown AI recruitment technology when it could find no way to stop its algorithm from systematically downgrading women's profiles for technical jobs and deducing its ground truth – all ideal candidates were men (Lavanchy, 2018).[13] Like any other algorithm, it was trained to observe patterns in large datasets as a way of predicting outcomes. It reviewed all the CVs that its employees had originally submitted to Amazon over a ten-year period to construct a profile of the 'optimal worker'. Given the low numbers of women working in technical roles in Amazon, it was not altogether surprising that the algorithm compounded existing sex segregation patterns within the company; 'the algorithm quickly spotted male dominance and thought it was a factor in success'. That is to say, the algorithm constructed its own 'truth'. This was a case of obvious fault but a more subtle version of 'truth' may well have gone undetected.

Additional to intelligent screening software is 'digitised interviewing', whereby candidates respond to automated questioning and their answers are documented by their choice of words, speech patterns, facial expressions and the like (Drage and Mackereth, 2022). These responses are then assessed in terms of 'fit' for specific duties of the advertised role as well as the employing organisation's culture, including their 'willingness

---

[13] Simply abandoning the algorithm to return to human bias on which the algorithm was trained in the first place is unlikely to yield better results in this example.

to learn' and 'personal stability'.[14] On the upside, it is clear that a great deal of time is saved through using these technologies, especially in the case of large numbers of applicants. The argument can be made that any of these algorithmic biases are no worse than those of humans and that in fact, algorithms can even be set to 'ignore' any social characteristic such as ethnicity, gender and age. Irrespective, it is not difficult to see how the plurality of humanity that Arendt saw as so valuable to social transformation is diminished to a very small number of candidate profiles after several rounds of intelligent screening software, which, in fact, might be better known as 'consolidation of human bias software'. What hope is there for radical new ideas, perspectives and a change in thinking?

In a similar vein, O'Neil's work showed how algorithmic technology used in fields such as insurance, education, health and policing systematically disadvantages the poorer cohorts of society:

> Employers... are increasingly using credit scores to evaluate potential hires. Those who pay their bills promptly, the thinking goes, are more likely to show up to work on time and follow the rules. In fact, there are plenty of responsible people and good workers who suffer misfortune and see their credit scores fall. But the belief that bad credit correlates with bad job performance leaves those with low scores less likely to find work. Joblessness pushes them toward poverty, which further worsens their scores, making it even harder for them to land a job. It's a downward spiral. (2016: 17)[15]

This sort of 'evaluation' is what both Katz (2021) and Fraser (2021) mean when they discuss the provocative idea of the management of 'human waste' – those who are subjected to the increasing precarity of neoliberal societies and systematically 'expelled from any possibility of steady employment' (Fraser, 2021: 162). An illuminating example is the US case of *State* v. *Loomis* (2016),[16] in which an algorithmic recidivism risk

---

[14] See Hirevue promotional materials, e.g., at www.hirevue.com/blog/hiring/hirevue-hiring-intelligence.

[15] Also see, e.g., Eubanks (2018).

[16] *State of Wisconsin, Plaintiff-Respondent v. Eric L. Loomis, Defendant-Appellant* 881 N.W.2d 749 (2016). Report available at www.courts.ca.gov/documents/BTB24-2L-3.pdf.

assessment was employed by a Wisconsin criminal court to 'calculate the likelihood of an individual with the offender's background committing another crime based on an evaluation of actuarial data'. Loomis had been charged with five criminal counts related to a drive-by shooting in Wisconsin. Loomis denied being involved but admitted to having driven the relevant car later in the same day. On these grounds, he pleaded guilty to two of the five charges – 'attempting to flee a traffic officer and operating a motor vehicle without the owner's consent'.[17] However, based largely on an algorithmic recidivism assessment entitled 'Correctional Offender Management Profiling for Alternative Sanctions' (COMPAS) (Brennan et al., 2009), Loomis was sentenced to six years' imprisonment and a further five years' extended supervision.[18] The precise methodological approach of COMPAS's predictive model, based on a questionnaire and aggregate data, was deemed to be market-sensitive and therefore only the results, devoid of algorithmic detail, were reported to the court. Consequently, Loomis appealed on the grounds that the reasons for his particular sentence were known neither to him nor the court. His appeal was rejected as was a subsequent appeal to the Wisconsin Supreme Court (Harvard Law Review, 2017). The usage of this sort of AI for the determination of criminal sentences goes against the very essence of what it means to be 'innocent until proven guilty'. The profile of those convicted before Loomis inadvertently foretold his fate, as did his for future defendants. The algorithmic recidivism assessment, COMPAS, would never afford Loomis and others identified to be 'like him' the possibility of having behaved in ways that were unexpected. The algorithmic decision had put that 'beyond doubt'.[19]

If an algorithm, such as COMPAS, is used to predict the probability of future criminal behaviour based on measurable proxies such as being arrested (Courtland, 2018), and AI-recommendation systems select new candidates based on the profiles of previously selected candidates, it is no

---

[17] *State* v. *Loomis* at 754.

[18] In the state of Wisconsin, extended supervision refers to a set period of time beyond a prison sentence whereby the convicted individual is required to adhere to a certain set of behavioural conditions which are regularly reviewed by a parole officer. Failure to meet these requirements results in a return to prison.

[19] Amoore (2019:149).

wonder that AI can significantly contribute to exacerbation of stereotypes, biases and other fault-based injustices. Scholars such as Noble (2018), Buolamwini and Gebru (2018) and Benjamin (2019) have also shown the far-reaching racialised dimensions of AI and the reduction of racialised peoples to crude stereotypes and misidentifications. Like other technologies and scientific practices, AI plays a central role in the production and codification of racialised understandings of the human body (Felt et al., 2017). Benjamin, for example, describes how Google users searching for 'three Black teenagers' were presented with 'criminal mug shots' (2019: 93). Similarly, Google Vision Cloud identified images of Black individuals holding a thermometer as holding a 'gun' (Kayser-Bril, 2020). With an estimated 5.6 billion searches a day across the globe, the Google algorithm is used by hundreds of millions of people daily to understand, and explore virtually, the world in which they live. The scale of influence is unfathomable. While tech companies might argue that their algorithms only rank and reflect back amassed human views and actions, some algorithmic products clearly serve to choreograph thought.[20] In the documentary, *The Social Dilemma* (2020), Aza Raskin (former head of user experience at Mozilla) explains, 'two billion people will have had a thought that they did not intend to have because a designer at Google said this is how notifications work on that screen that you wake up to in the morning'.

There is of course an argument that we humans have never before been exposed to so much plurality of thought, culture and experiences through the wonders of global social media platforms, where millions of us spend hours every day asking questions and sharing our opinions with strangers around the world through X (formerly Twitter), Facebook,

[20] In part, the AI sector is at fault for its inability to detect discrimination, a fact mirrored by its lack of workforce diversity (Hayles, 2023). On this point, Costanza-Chock argues that to diversify the technology workforce is of course a good idea and will go some way to improving AI-generated harm, but 'it will not automatically produce a more diverse default imagined user' that in turn informs the design of AI and the assumptions built into a system's ground truth. Rather, it is the case that 'unless gender identity, sexual orientation, race/ethnicity, age, nationality, language, immigration status and other aspects of user identity are explicitly specified, even diverse design teams tend to default to imagined users who belong to the dominant social group' (2023: 372). This is a topic I return to in Chapter 4.

Instagram, WeChat and the like. Some estimates report that about half the world's population use social media in some form (Chaffey, 2022). However, the primary objective of social media companies is to encourage us to forge new common habits, not to improve our lives per se but to behave in ways that fit a particular business model that attracts more and more users for longer and longer periods, which produces ever greater amounts of data that tech companies can use and sell. As tech investor Roger McNamee explains, everyone in your newsfeed seems just like you, and when you are surrounded by what you think are like-minded people from all over the world, 'it turns out when you are in that state, you are highly manipulable' (2020: np). What is more, the new wave of AI – generative AI – is set to compound this problem a great deal further. Of particular concern is the capacity of generative AI to create disinformation on a vast scale. Take, for example, recent fake images produced on MidJourney, which look convincingly like the US 1969 Apollo 11 spaceflight (in which Commander Neil Armstrong and lunar module pilot Buzz Aldrin were the first ever humans to land on the Moon) was in fact constructed on a film set.[21] Conspiracies around the Apollo missions are still among the most widely repeated in the world despite numerous evidence-based rebuttals (Runciman, 2020), indicating the appeal of disinformation. As Tristan Harris, former design ethicist at Google, explains, 'fake news spreads six times faster than true news' on social media platforms (2020: np).

As we can see from these examples, a complex picture emerges of how AI exacerbates injustice. AI-generated injustices, such as algorithms that propagate discriminatory stereotypes or churn out fake news, can be traced back to culpable tech companies, designers and, in some cases, users. These are questions of liability. However, while it is clear that there is nothing neutral about the algorithms of AI, how could we conceivably trace the complex formation of individuals' thoughts and beliefs back through the myriad of feeds seen and engaged with across a population over time? At some point the effects of these interactions operate at the structural level and are untraceable as discussed in Chapter 1. Benjamin's work on the automation of racial discrimination, for

---

[21] See the Royal Museums Greenwich website on this example, at www.rmg.co.uk/stories/topics/moon-landing-conspiracy-theories-debunked.

example, helps us to see how AI-generated injustice not only manifests in forms that are traceable to liable agents of fault but also operates on the macro-structural scale:

> By pulling back the curtain and drawing attention to forms of coded inequity, not only do we become more aware of the social dimensions of technology but we can work together against the emergence of a digital caste system that relies on our naivety when it comes to the neutrality of technology. This problem extends beyond obvious forms of criminalization and surveillance. It includes an elaborate social and technical apparatus that governs all areas of life (2019: 11).

Here I see a link between Benjamin's reach beyond 'criminalization and surveillance' to 'apparatus that governs all areas of life' and Young's reach beyond 'fault-based injustice' to structural dynamics whereby many societal outcomes are determined by macro-relational forces. As I have attempted to explain, while it is vital for the pursuit of justice to act upon traceable actions that cause direct or indirect discrimination or harm, these do not explain the full extent of injustice. Structural injustice, greatly aided by 'an elaborate social and technical apparatus', is generated by the everyday structural actions of ordinary people going about their legitimate pursuits of private interest. The fact that so many of us are 'thoughtless' in our dependence on AI is a structural dilemma.

The level of causal complexity between individual action and structural injustice is so intricate and convoluted on a mass scale that it is not meaningfully traceable with any existing epistemological tools. For example, in what meaningful ways could we trace all the individuals who are implicated in the global demand for general technologies which result in increased international AI R&D that, in turn, produce new social media products, some of which may create unintended negative consequences for particular groups? I think Young's view would have been that although there may be some traceable elements of fault in this example such as some discriminatory algorithms (which ought to be addressed through claims of liability), it is more productive for macro-level change to speculate about what can be done to reset the background conditions of this sort of structural injustice in which Tech Industry interests override those of the public. How to think about governing AI with a particular focus on its imbrication within social power structures is a theme I shall return to, but first I want to explore how the traditional liability measures for addressing AI-generated harm are becoming ever more inadequate as algorithmic decision-making becomes ever more opaque to us.

AI, THE BLACK BOX AND UNTRACEABILITY

We build amazing models, but we do not quite understand them. And every year, this gap is going to get a bit larger.

Jason Yosinski, AI scientist

Opacity is key to the AI sector. Not only does it give AI technology an air of rational neutrality, but also no tech company is keen to share the inner workings of its algorithms in such a competitive market. However, beyond the protection of trade secrets is a different sort of opacity – often referred to as a lack of AI 'explainability' or 'interpretability':

No one really knows how the most advanced algorithms do what they do. (Knight, 2017: np)

The training procedure in Deep Learning determines the settings of millions of parameters which interact in a complex way. It is extremely difficult to reverse engineer a Neural Network. We have reached a point where even the creator of an algorithm does not understand it completely. (Nielly, 2020: np)

The highly complex workings of deep machine learning are increasingly designed to develop independently of humans, described earlier as 'unsupervised learning'. These processes lead to a form of self-reproduction in new and unpredictable ways that are unfathomably dense and complicated to humans (Castelvecchi, 2016; von Eschenbach, 2021). While we might be able to understand something of the data inputs and outputs, the algorithmic decisions themselves, even with the help of other algorithms, are often untraceable. This is what is meant by the 'black box' of neural networks (Nielly, 2020), and is becoming not so much a question of ignorance but rather one of impossibility. This realisation is one that chimes with Young's account of the untraceability of structural injustice – the causal chain between decisions and outputs is undetectable.

Ironically, as we give or leak more and more of our data to what Lawrence (2015) has called a system of 'digital servitude' to the institutions, firms and governments that hold and utilise it, we become more transparent and controllable as the algorithms become more opaque and independent. This is what Runciman (2023) refers to as 'the handover'.

Writing soon after Turing posed the famous question, 'Can machines think?' (1950: 433), Arendt recognised that automation and AI were seen as crucial in accelerating societal progress, but was unconvinced that these new technologies ought to be thought of as the primary drivers of prosperity. Machines and the 'advent of automation' were seen by Arendt as having great potential for the atrophy of critical thought. Such thoughtlessness rendered the individual no more than an aptly described homogenous 'automaton' (1998 [1958]: 126). One of Arendt's (highly controversial)[22] influences and a foundational scholar of the sub-field of philosophy of technology, Heidegger (1977: 4) warned of assuming technology as a neutral force. He vehemently argued that we should not see it as a means to an end, something of simply instrumental value as it is often promoted, but rather as fundamentally changing the way we conceive of and act upon our social and natural environment. Heidegger argued that while recognising what he saw as the inevitable rise of technology, we ought to resist its domination of human life by actively enabling other sorts of interpretations of our environment so as to give us the capacity to reflect and judge – described as 'releasement' from technology, a key feature of which is the capacity to question and critique (1977: 35).

Nevertheless, the imperative to rely on algorithms has become all too compelling (and with little opportunity of 'releasement'). As Floridi (2014, 2020) has argued, algorithms are now core to what most people think of as human well-being (medical predictions, environmental planning, transport logistics, complex financial data management, etc.). In

---

[22] The philosopher Martin Heidegger, who later became a Nazi Party member, formed a relationship with his student Hannah Arendt in the mid-1920s (see Benhabib, 2021).

this sense, Arendt (1998 [1958]: 3) was sceptical about humanity's capacity to think: 'it could be that we ... will forever be unable to understand, that is, to think and speak about the things which nevertheless we are able to do'. By increasingly deferring to algorithmic decisions in every sphere of life, not only are we relinquishing or avoiding accountability, but also, most importantly, we are diminishing our capacity to reflect on and judge our norms and assumptions. Indeed, technologies that threaten humanity's plurality of thought are a threat to the potential for politics itself. Delacroix argues that much like muscle atrophy that comes from lack of physical activity, '[r]eliance upon non-ambiguous systems – whose opaque, multi-objective optimisation processes makes any effort of critical engagement redundant – will affect the extent to which we are made to flex our "normative muscles" in the longer term' (2021: 12–13). These are vital themes I return to in Chapter 4.

Both traceable liability-based as well as the broader background conditions of structural injustice are perpetrated by AI. Both forms of injustice bring with them an increasing thoughtlessness for those reliant on AI and we have seen that trend intensified by the growing tendency of many powerful decision-making algorithms to be untraceable. However, as well as our increasing reliance on algorithms to think for us, there is also the question of human labour replacement by AI and its capacity to exacerbate structural injustice in other ways.

AI AND HUMAN LABOUR: A HUNDRED YEARS TO BLISS?

John Maynard Keynes famously warned in 1930 of 'technological unemployment' (1951 [1930]: 364).[23] For Keynes this was to be but a 'temporary phase of maladjustment' during which time technology would replace human labour to the detriment of the masses in the short to mid-term but that in the longer term, it would in fact 'solve mankind's

---

[23] Keynes's article entitled 'Economic possibilities for our grandchildren' started out as a lecture to students of Cambridge University. Keynes was worried that students, disillusioned with social and economic inequality, would be allured by communism. He sought to convince them that if they were to hang on, new technologies would bring an end to material need.

economic problems'. Indeed, Keynes predicted that within a hundred years of his writing, the UK's population could reach 'our destination of economic bliss' by 2030 (373) whereby 'technical efficiency' (358) would rise to such an extent that 'absolute' material needs would be satisfied with the minimal amount of labour effort: 'There will be ever larger and larger classes and groups of people from whom problems of economic necessity have been practically removed' (373). Keynes very much welcomed this future; 'I look forward, therefore, in days not so very remote, to the greatest change which has ever occurred in the material environment of life for human beings in the aggregate' (372).[24]

More recently, Winchester (2023) argues that an algorithmic revolution that brings about technological unemployment has the potential to release humans from what he sees as all the unnecessary knowledge that we must carry around in our minds to facilitate the labour that could be performed by AI. He sees this as freeing the potential for new forms of human genius in the future.

The seminal study on technological unemployment was first published in 2013 by Frey and Osborne. They employed an occupation-based methodology and predicted that 47 per cent of jobs in the United States were susceptible to automation in the near future. Subsequent authors such as Arntz et al. (2016) were highly critical of this approach and advanced an alternative task-based methodology (i.e., focusing on the different tasks performed within an occupation that could be divided up between a human worker and robotics/AI, thereby maintaining some function for humans). Using this alternative methodological approach, Arntz et al. estimated that only 9 per cent of jobs in the US labour market would become susceptible (2016: 4). Other studies using task-based methodologies such as that of PwC (2022) produce figures somewhere in the middle, predicting that by 2030 as many as 38 per cent of the US labour force[25] and 30 per cent of the

---

[24] See Skidelsky and Skidelsky (2013) for an excellent discussion of Keynes's GDP predictions, and interpretation of material wants and the good life.

[25] Current workforce in the United States is 158.7 million (US Bureau of Labour Statistics, 2023).

UK's[26] will be at risk of AI/robotic replacement. Similarly so in Germany, at 36 per cent.[27] Widely regarded as the most comprehensive predictive study to date, PwC's report argues that automation will boost productivity and increase wealth which in turn will be spent or invested. In doing so, the report concludes, AI and robotics will generate new jobs elsewhere in the economy, particularly in the service sectors identified as specifically dependent on human capabilities that are difficult to automate. This perspective is supported by the World Economic Forum report (2020), which predicts the decline of jobs in sectors such as mining and metals and an upsurge in professional services, resulting in an optimistic prediction of concentrated net job gain.[28]

In terms of productivity gains, those sectors with the most opportunity to substitute human labour are predicted to be those 'likely to see the largest productivity gains from AI' (PwC, 2017: 5). In Germany, *Bild* – the largest selling newspaper in Europe – offers a stark, recent example of this practice. In February 2023, Mathias Döpfner, the CEO of Springer-Verlag which owns *Bild*, announced that due to the 'opportunities of artificial intelligence' there would be 200 redundancies resulting in an £85 million saving on labour costs. It was explained that *Bild* would 'unfortunately be parting ways with colleagues who have tasks that in the digital world are performed by AI and/or automated processes' (Hanfeld, 2023).[29] Andy Haldane (former Chief Economist, Bank of England) has argued, for example, that AI will result in an extensive 'hollowing out' of the labour market, creating increasing inequality and a greater struggle for many people to make a living (Haldane, 2015).[30] A recent survey of 1,031 UK workers undertaken by a AI-recruitment company reported that 'a third of women (33%) and over two-fifths

---

[26] Current workforce in the UK is 33 million (ONS, 2023a).

[27] Current workforce in Germany is 45.7 million (Destatis, 2023).

[28] A similar picture emerges from the UK Office for National Statistics (ONS, 2019b).

[29] Another recent example is that of BT which has announced 55,000 jobs will be cut by 2030 with AI substituting a large number of roles (Sillars, 2023).

[30] Haldane argued in a speech to the Trades Union Congress in 2015, and on BBC Radio 4 *Today* (BBC, 2018a), that this industrial revolution is different from previous ones due to the capacity of AI to substitute not only human manual capabilities but cognitive ones also.

(43%) of men said that it was likely or very likely that their current job role could be replaced by technology or machines (such as smart software, automation or robotics) in the future' (CIPHR, 2022).[31]

With the rise of zero-hour contracts,[32] economist and tech investor Bill Janeway (2018: np) argues that 'it is not hard to see how the digitally enabled gig economy could become the ultimate realization of Marx's "reserve army of labor" – fully commodified human beings, available to capitalists on demand'. Under these increasingly precarious conditions, low- to mid-skilled labour could be reduced to ever lower-paid tasks that cost less than the costs of automation. On this particular point, Marx's observations translate directly to contemporary concerns:

> The Yankees have invented a stone-making machine. The English do not make use of it because the 'wretch' who does this work gets paid for such a small portion of his labour that machinery would increase the cost of production to the capitalist. In England women are still occasionally used instead of horses for hauling barges, because the labour required to produce horses and machines is an accurately known quantity, while that required to maintain the women of the surplus population is beneath all calculation. Hence, we nowhere find a more shameless squandering of human-labour power for despicable purposes than in England, the land of machinery. (1990 [1867]: 516)

Harari's (2017, 2021) view is that the new AI age will create 'the useless class' of humans as a consequence of diminishing demand for human labour. He argues that the impact of AI will not be contained in one single event 'after which the job market and the educational system will settle into a new equilibrium' but instead it will be 'a cascade of ever-bigger disruptions' (2017: 325).

Computer scientist, Melanie Mitchell (2019: np), argues that those who invoke the spectre of the superintelligent being humanity, whom

---

[31] See also the latest report from Goldman Sachs (2023) on AI replacement of human labour in the United States.

[32] Before Covid-19 took hold of the UK economy in 2020, more than a million people were employed in the UK labour market on zero-hours contracts, whereby employers guarantee no fixed hours and employees correspondingly lack core employment rights (ONS, 2021).

she calls the 'superintelligentsia', are enthralled by science fiction, not reality.[33] To put a finer point on it, the EU Commissioner for Competition, Margrethe Vestager, said that '[p]robably [the risk of extinction] may exist, but I think the likelihood is quite small. I think the AI risks are more that people will be discriminated [against], they will not be seen as who they are' (quoted in Milmo and Hern, 2023). This brings us back to the claims of AI's partiality by Amoore (2019) and a sense of how on a mass scale, structural injustices could develop in addition to and at the more convoluted complex scale, beyond traceable liability.

Stephen Hawking, however, takes us a step further. He warned that AI will soon reach a point where it will not only supersede human intelligence but increase its intelligence at an exponential rate so that it will operate well beyond our control (BBC, 2014). As Barrat (2013: np)[34] reminds us, 'humans only rule the future not because we are the fastest or strongest creature but because we are the most intelligent', and when we share the planet with something more intelligent, it will rule the future. Whatever turns out to be true, one thing that seems certain is that, contra Keynes's predictions, AI is not brining the average worker 'economic bliss. Mitchell and Brynjolfsson's (2017) seminal research on the US economy demonstrated 'how technology has enabled industry to become more productive without paying the median worker more'. So, while productivity levels hit an all-time high in the United States for example, income for the bottom 50 per cent of earners has stagnated since 1999.[35] And in the absence of relevant policy responses, Mitchell and Brynjolfsson conclude that policymakers are 'flying blind' into the AI age.

Orienting AI towards the public interest is one of the most important issues of our time. This begs the question of what sort of approach states have taken to mitigate against AI-generated harm, including not only

---

[33] Similarly, Andrew Ng, Chief Scientist at Baidu, the leading Chinese search engine tech company, famously suggested that 'Worrying about sentient AI is like worrying about overpopulation on Mars' (Williams 2015).

[34] Interview with James Barratt on the *Today Programme* – 19 April 2023 – BBC Radio 4. See also Barratt (2013).

[35] See also Bivens and Mishel (2021).

those liability-based harms that I have described but also those that operate at the structural level. I will now consider this question through a focus on the UK, which considers itself to be at the 'forefront of safe and responsible AI' (HM Government, 2021b) and a 'global AI super-power' (HM Government, 2022a), by way of example.

AI GOVERNANCE: THE EXAMPLE OF THE UK

The UK Government predicts that annual investment by UK companies in AI technologies will rise dramatically to an estimated £83.5 billion in 2040 (HM Government, 2022b). Current thinking on how to manage such a shift in the UK's economy is best captured by the 2021 UK National AI Strategy (HM Government, 2021c), the 2022 policy paper 'Establishing a pro-innovation approach to regulating AI' (HM Government, 2022a) and the 2023 White Paper 'A pro-innovation approach to regulating AI':[36]

> Our approach relies on collaboration between government, regulators and business … we do not intend to introduce new legislation. By rushing to legislate too early, we would risk placing undue burdens on businesses. (HM Government, 2023a: np)

The thrust of the 'new approach to regulating AI' is a liability-based approach with a focus on 'where there is clear evidence of real risk … rather than hypothetical or low risks associated with AI. We want to encourage innovation and avoid placing unnecessary barriers in the way'. (HM Government, 2022a: np).The current position is a non-statutory approach with 'lighter touch options, such as guidance or voluntary measures' and little opportunity for scrutiny (2022a: np).

Some elements of the UK's liability approach to AI governance are borrowed from the EU in the form of the 2016 General Data Protection Regulation (GDPR), which was not written specifically with AI in mind

---

[36] See also the UK's Science and Technology Framework setting out its plans for £20 billion of R&D investment (HM Government, 2023b). This pro-innovation approach mirrors something of the US Algorithmic Accountability Act 2022, which requires companies to self-assess the impact of its AI development and usage.

but provides a general right to the protection of personal data.[37] Beyond the UK, the EU has gone further still with what has been donned 'the world's first rules on Artificial Intelligence' and the European Commission's embryonic AI Act 2021 – a risk-based regulatory approach that is specifically designed not to impede AI markets by restricting regulation to 'those concrete situations where there is a justified cause for concern' (European Commission, 2021: sec 1.1.1).[38] The Act, set to come into force in 2026, requires, for example, that 'high-risk AI'[39] adhere to a strict set of restrictions in the interest of the public which the Act defines as the 'health and safety, protection of fundamental rights, democracy and rule of law and the environment, as recognised and protected by Union law' (Rec 5 and 13). Here the focus is on the protection of individuals' privacy, safety and security; transparency and fairness; non-discriminatory programmes; and professional responsibility for AI design. Providers of high-risk AI must devise systems of risk management that can be 'known and ... reasonably foreseeable risks that the high-risk AI system can pose to the health or safety of natural persons, their fundamental rights including equal access and opportunities, democracy and rule of law or the environment when the high-risk AI system is used in accordance with its intended purpose and under

[37] Further GDPR details are available on the UK Government website: www.legislation.gov.uk/eur/2016/679/contents.
[38] The June 2023 Amendments to the EU AI Act can be found here www.europarl.europa.eu/doceo/document/TA-9-2023-0236_EN.html.
[39] The weight of the Act focuses on 'high risk AI' which is classified into five categories (JDSUPRA, 2021): 'Biometric identification and categorisation: AI systems intended to be used for real-time and post remote (e.g., CCTV footage) biometric identification of individuals; Management and operation of critical infrastructure: AI systems intended to be used as safety components in the management and operation of road traffic and supply of water, gas, heating, and electricity; Education and vocational training: AI systems intended to be used for determining access or assigning individuals to educational and vocational training institutions, or for the purpose of assessing students in educational institutions; Employment, workers management and access to self-employment: AI systems intended to be used for recruitment or selection of individuals, notably, advertising vacancies, or for making decisions on monitoring and evaluating workplace performance; Access to and enjoyment of essential private services and public services and benefits: AI systems intended to be used by or on behalf of public authorities to evaluate individuals for public assistance benefits and services, or to evaluate the creditworthiness of individuals.'

conditions of reasonably foreseeable misuse' (Art 9.2a) and 'is reasonably judged to be acceptable, provided that the high-risk AI system is used in accordance with its intended purpose or under conditions of reasonably foreseeable misuse. Those residual risks and the reasoned judgements made shall be communicated to the deployer'. (Art 9.4). As with the UK's approach, however, there is no real guidance on how these risks should be interpreted in the increasing number of cases where AI impacts on individuals, groups or institutions. The protection generated by this legislation is limited to that which is only traceably liable, 'known' and 'acceptable' or else left to the realm of generic voluntary ethical codes.[40] Neither approach is effective for structural sorts of harm. Helberger and Diakopoulos (2023: 2) argue, for example, that generative AI systems like ChatGPT 'are not built for a specific *context* or conditions of use, and their openness and ease of control allow for unprecedented *scale* of use'. This means that generative AI would be very difficult to categorise as 'high' under the EU AI Act due to very little consistency in how providers (those liable) will use the algorithm and therefore great unpredictability of the potential risks.[41] As Helberger and Diakopoulos explain:

> With 100 million active users in the first months after its launch, ChatGPT has been described as the 'fastest-growing consumer application ever launched.' The pure scale of adoption, in combination with the versatility and general purpose characteristics of the technology, challenge the AI Act's risk-based approach ... it is simply impossible to predict ... what the risks are that we can expect from unleashing extremely powerful AI models on society. (2023: 4)

---

[40] There are however, some bold new measures intended for the final version of the Act reported by the *Guardian* (see O'Carroll, 2023), such as an attempt to address copyright infringement of the work of academics, artists, musicians, journalists, etc. whose work has been scraped from the internet and used to train various forms of generative AI such as ChatGPT without any citations or royalties. The Act will require that all training data is registered and that legal requirements as to the use of that data are adhered to.

[41] Also see Anderson (2023).

Thinking back to my discussion of structural responsibility in Chapter 1, the minimalist approach to managing the risks of AI are limited to 'evidence of real risk' and very much in line with the liability model of political responsibility. Seeking out who is at fault for the outcomes of crudely designed algorithms would fall squarely under liability (whether legal or moral, indirect or direct). There is, however, no opportunity for wider public speculation on the structural dynamics of algorithmic decision-making and our increasing societal thoughtlessness. Rather, 'AI practitioners and disruptors' are called forward to think about what sort of policy, if any, might be best for us all with a heavy focus on facilitating AI sector growth (HM Government 2022a: np).

The UK's approach to governing AI is grounded in the research and thinking done by the House of Lords Select Committee Report, 'AI in the UK: Ready, willing and able?' (2018). One of the key points recommended by the House of Lords, and very much welcomed by the Government, was a rejection of the need for a central regulatory public body with responsibility for AI: 'Blanket AI-specific regulation ... would be inappropriate' (House of Lords Select Committee, 2018: 386). This position sits in line with the Government's Industrial Strategy: 'to work with businesses to develop an agile approach to regulation that promotes innovation and the growth of new sectors' (HM Government, 2018: 103) and was restated in Lord Evans of Weardale's AI and Public Standards Report to the UK Government: 'the UK does not need a new regulator' (Evans, 2020: 47). Naturally, the drive for innovation is supported by a number of the tech giants. In written evidence to the Select Committee, Google sought to 'encourage a cautious and nuanced regulatory approach that will allow innovative uses to flourish and reach their full potential' (written evidence, 7.3: 627–629),[42] and Microsoft articulated that 'policy discussion should prioritise broad development and deployment of AI across different sectors and continued AI innovation' (written evidence, 33: 1016).

---

[42] Google's written evidence may be found at: www.parliament.uk/business/lords/media-centre/house-of-lords-media-notices/house-of-lords-media-notices-2017/october-2017/lords-artificial-intelligence-committee-publishes-written-evidence/.

In line with many states, however, the House of Lords did recommend a voluntary AI Code of Ethics in place of more comprehensive regulation. This code consists of five major elements common to many national and international ethical codes:[43]

(1) Artificial intelligence should be developed for the common good and benefit of humanity.
(2) Artificial intelligence should operate on principles of intelligibility and fairness.
(3) Artificial intelligence should not be used to diminish the data rights or privacy of individuals, families or communities.
(4) All citizens have the right to be educated to enable them to flourish mentally, emotionally and economically alongside artificial intelligence.
(5) The autonomous power to hurt, destroy or deceive human beings should never be vested in artificial intelligence.

(House of Lords Select Committee, 2018: ch 9, 417)

How effective are these non-binding ethical codes, and are they sufficient for the structural dynamics of AI? I'll consider them in reverse order.

○ *The Autonomous Power to Hurt, Destroy or Deceive Human Beings Should Never Be Vested in Artificial Intelligence*

In December 2021, UN member state representatives met in Geneva to discuss the banning of autonomous weaponry (UN, 2021b). However, the UK and the other permanent members of the Security Council refused to vote for a ban.

Despite making what had meant to be a clarificatory statement in 2017 – 'the UK does not possess fully autonomous weapon systems and has no intention of developing them' (Ministry of Defence, 2017: 14) – in May 2018 the UK's MoD launched a US–UK AI defence collaboration to 'harness new technologies and approaches to stay ahead of our adversaries and keep safe'. The new US–UK military AI hub is designed

---

[43] As Jobin et al. (2019) demonstrate in their study of eighty-four national and international ethical frameworks, common generic themes tend to feature throughout such frameworks (see also Browne et al., 2023).

to encourage 'our best and brightest to develop new capabilities in everything from Artificial Intelligence and autonomous weapons to advanced cyber and robotics' (Williamson, 2018).[44] Similarly, in 2020 the Government announced that its Defence Science and Technology Laboratory was 'looking at how future defence platforms can be designed and optimised to exploit current and future advances in automation, autonomy, machine learning and artificial intelligence' (HM Government, 2020a). Whatever one thinks of military exceptions to the rule, ethical codes have not limited the development of lethal autonomous weapon systems defined as any 'system that decides and acts to accomplish desired goals, within defined parameters, based on acquired knowledge and an evolving situational awareness, following an optimal but potentially unpredictable course of action' (Department for Digital, Culture, Media and Sport, 2021: 14). The Government has responded by setting up a new Defence Centre for AI Research (DCAR) as part of the newly released MoD Defence AI Strategy (HM Government, 2022c) to accelerate the research, development, testing, integration and deployment of world-leading AI. Within this strategy there is a new pledge: 'there must be context-appropriate human involvement in weapons which identify, select and attack targets [lethal autonomous weapons]. This could mean some form of real-time human supervision, or control exercised through the setting of a system's operational parameters' (Ministry of Defence, 2022: 3). This presumably makes the human 'in the loop' or the 'human on the loop'[45] liable for any algorithmic judgement, but it is not explained in what ways and to what extent that human is liable. As Amoore (2019) points out, past

---

[44] Note the different definitions here – 'the UK does not possess *fully* autonomous weapon systems' [emphasis added] versus 'to develop new capabilities . . . autonomous weapons'. This is significant because until 2022 the UK used this definition of fully autonomous weapons to describe a technology that does not exist (i.e. where there is no human element to the ground truth). It is therefore true to say 'we do not possess something that does not exist'. See, e.g., House of Lords Library (2020).

[45] As Coco (2023: 3) explains 'a "human in the loop" setting is one in which the machine has identified a target and the human must validate it before the machine initiates engagement. Whereas, a "human on the loop" setting is one in which the machine identifies the target and commences engagement unless the human overrides the machine's determination and halts the engagement'.

human decisions are lodged within the algorithm as part of ground truth, and in this sense the human has always been 'in the loop' of algorithmic decision-making. How would an additional human in the loop be liable? Amoore's work shows us how practitioners who work with AI are changing their perceptions of what can be known or done, but this raises questions as to how blameworthy they can be for all the human actions that have accumulatively manifested in the ground truth of a particular system and its corresponding algorithmic decisions.

A cruder point can be made: given that AI currently fails to distinguish a chihuahua from a blueberry muffin,[46] what would the additional 'human in the loop' be liable for if a lethal autonomous weapon annihilated the wrong target? It is not at all clear what is being achieved by the 'human in the loop' in such a case other than to create a superficial agent of liability. This seems to me to be a case of scapegoat ethics in a context where the prospect of large-scale error is paramount: indeed it would be prudent to remember that, 'the difference between human error and algorithmic error is like the difference between mailing a letter and tweeting' to the masses (Dawes, 2021).[47]

These examples demonstrate that even in terms of liability, the 'liable human' cannot address the potential injustice of autonomous weapons, let alone the macro-power dynamics that threaten us all on a structural level. It is also noteworthy that this particular ethical principle – '[t]hat autonomous power to hurt, destroy or deceive human beings should never be vested in artificial intelligence' – was dropped from the 2022 Defence AI Strategy[48] and the 2023 AI Regulation White Paper

---

[46] See, e.g., Minds at Work (2018).

[47] To give a recent example, a 2021 study found that the majority of 'hate tweets' posted about Meghan Markle (about 70 per cent) were generated by just 83 Twitter accounts and yet reached 17 million Twitter users (Davis, 2022).

[48] In the UK's current approach to the governance of AI, the House of Lords Ethical Codes were dropped and, in their place, the more generic 2023 White Paper AI principles were adopted. These are: (1) Safety, security and robustness; (2) Transparency and 'explainability'; (3) Fairness; (4) Accountability and governance; (5) Contestability and redress. These loosely align with the OECD (2019) AI principles.

As an aside, the MoD has its own ethical board which is inexplicably populated almost entirely by the medical profession; see www.gov.uk/government/groups/ministry-of-defence-research-ethics-committees#ethics-committee-members.

(Office for Artificial Intelligence 2023) on the pro-innovation approach to AI governance which was formulated without any substantive public consultation, an issue I shall discuss further in Chapter 4.

○ *All citizens have the right to be educated to enable them to flourish mentally, emotionally and economically alongside artificial intelligence*

In the face of disruptions to the labour market, many of the least educated are at the highest risk of AI replacement and the macro-structural dynamics that would ensue from high rates of technological unemployment.[49] With something of Keynes's optimism, however, the UK Government's general stance is that automation will free up individuals to invest in upskilling and has provided £117 million to fund PhDs for AI researchers (UKRI 2024). Of course, researching at the level of PhD caters to a tiny proportion of the future labour force. What can those with less specialist training expect in terms of new employment opportunities? The Covid-19 pandemic had such an impact on the labour market that it is impossible to present a counterfactual. However, just before the pandemic in 2019, we can see clearly that despite marginal economic growth in the UK, real wages were lower than before the 2008 recession (ONS, 2019a). Hence while there were technically more people in work (with unemployment down to a record 3.85 per cent), low- to mid-skilled workers tended to be in lower-paid and more insecure jobs. Indeed, of the million pre-pandemic workers employed on zero-hour contracts in the UK, as many as 73 per cent experienced precarious employment with less than twenty-four hours' notice (TUC, 2020). Peyton-Jones (2024) suggests that real wages are likely not to return to 2009 levels for another ten years.

Even if there were to be a huge rebirth of the labour force through the digital economy, given that the jobs of low-skilled workers are most susceptible to automation, the concern must be that the UK's labour market policies merely expose these workers to greatly increased insecurity. Peyton-Jones (2024) explains that 'if recent UK trends continue, the proportion of national income accounted for by the highest 0.1% of

---

[49] PwC (2022) estimate that across twenty-nine OECD countries, up to 45 per cent of workers with a low level of education will be replaced by AI by 2030.

earners will increase from 5% to 14% by 2030'. Certainly the 'click economy' is not likely to bring a great many well-paid jobs for those displaced (Hagendroff, 2020), especially as much of the new educational resources for upskilling are focused on technical subjects. The work of Jones (2021) shows that the vast cohort of data taggers that work for the Mechanical Turk or directly for tech companies are often paid task by task at very low rates, some less than $2 an hour.

Promises of retraining for the AI age are undermined yet further as this ethical principle too has been dropped from the UK Government's newest version of the pro-innovation approach to regulating AI. This serves as another example of a widening vacuum of political responsibility, not only for the direct individual cases of precarity due to technological unemployment but also for the wider structural consequences.

○ *Artificial intelligence should not be used to diminish the data rights or privacy of individuals, families or communities*

This concern dominates discussions of AI and its relationship to the public interest. It is one area in which the UK Government has introduced legislation. The Information Commissioner's Office (ICO), the UK's data protection authority, works through the Data Protection Act 2018, which in turn channels the EU's GDPR.[50] These approaches are designed to enable individuals to know when automated decision-making processes or profiling (to predict behaviour or interests) are conducted using held personal data.[51]

There are two main problems with this approach, however. The first is that it is firmly directed towards the rights of individuals and thus

---

[50] For a full copy of the EU GDPR, see https://gdpr-info.eu/.

[51] See Data Protection Act 2018, s 14(4):

'Where a controller [the natural or legal person, public authority, agency or other body which, alone or jointly with others, determines the purposes and means of the processing of personal data] takes a qualifying significant decision in relation to a data subject based solely on automated processing – (a) the controller must, as soon as reasonably practicable, notify the data subject in writing that a decision has been taken based solely on automated processing, and (b) the data subject may, before the end of the period of 1 month beginning with receipt of the notification, request the controller to – (i) reconsider the decision, or (ii) take a new decision that is not based solely on automated processing.'

downplays the possible danger that, as the French mathematician and politician Cédric Villani puts it, 'legislation, which focuses on the protection of the individual, is not consistent with the logic introduced by these [AI] systems – i.e. the analysis of a considerable quantity of information for the purpose of identifying hidden trends and behaviour – and their effect on groups of individuals' (2018: 14). As discussed earlier, AI usually determines its outputs by observing and reinterpreting patterns across big datasets, and operates according to a particular ground truth. While these outputs will inform policy, recommendations or particular decisions, the algorithmic decision-making process is unlikely to be explainable in a way that could be traced to an individual case. Hence, such a right is extremely difficult to put into practice (HM Government, 2020b: 44). The second problem is that any protection that the Act affords is not likely to be experienced equally across society. Article 22 of the GDPR states that '[t]he Act includes the necessary safeguards such as the right to be informed of automated processing as soon as possible and also the right to challenge an automated decision made by a data controller or processor'. When one considers the ways in which algorithms impact our lives, influencing our access to information, to financial credit, culture, to employment and so on, it seems likely that those in greatest need of protection are also those least likely to access the remedies provided for in the Act. This brings us back to one of Young's central points that marginalised individuals and groups invariably find themselves at the sharp end of broad macro-level structural injustices with no power to effect change. This ethical issue links directly to the next element.

○ *Artificial intelligence should operate on principles of intelligibility and fairness*

The question of black-boxing was raised by the House of Lords Select Committee: 'We believe it is not acceptable to deploy any artificial intelligence system which could have a substantial impact on an individual's life, unless it can generate a full and satisfactory explanation for the decisions it will take' (2018: 105). In the French national AI strategy, a similar concern was raised: 'In the long term, the accountability of this technology is one of the conditions of its social acceptability. Regarding certain issues, it is even a question of principle: as a society we cannot allow certain important decisions to be taken without explanation' (Villani, 2018: 115). While acknowledging that businesses are reluctant to see their intellectual property divulged to third parties, the French policy nevertheless introduced a trade-off between business and public interest, in the form of an audit body of experts 'with the requisite skills ... [for the] auditing of algorithms and databases and for checking them using any means deemed necessary' (Villani, 2018). This concern, however, did not translate into a generic policy at the European level. In the new AI Act, only a subset of high-risk AI systems must make use of a third-party body – a 'notified body' – to conduct an independent audit.[52] This is because the Act follows a 'risk-based approach and imposes regulatory burdens only when an AI system is likely to pose high risks to fundamental rights and safety' (European Commission, 2021: EM 2:3: np). It is particularly noteworthy that the Act specifies 'For high-risk AI systems, the requirements of high-quality data, documentation and trace-ability, transparency, human oversight, accuracy and robustness, are strictly necessary to mitigate the risks to fundamental rights and safety posed by AI' (2021: EM 2:3: np). This sort of liability approach sounds promising where possible. However, as I have suggested, calls for 'trans-parency' carry increasingly little weight as deep machine learning becomes more powerful and opaque. Such algorithms continually evolve as they absorb more and more data, so that what might have been conceivably transparent milliseconds ago suddenly becomes opaque, and the possibility of a meaningful audit or traceable algorithmic decision-making process becomes less and less viable. This increasing limitation would seem to lend itself to the UK Government's position:

> overemphasis on transparency may be both a deterrent and in some cases such as deep learning prohibitively difficult. Such considerations need to be balanced against positive impacts use of AI brings ... [The matter of] informing the public of how and when AI is being used to make decisions about them and what implications this will have for them personally ... [is] left to individual businesses to decide on whether and in what way to inform consumers of AI's deployment. (HM Government, 2018: 59)

---

[52] See Edwards (2022).

Such a light-touch approach is firmly limited to questions of liability and does not attempt to address potential structural AI harms that operate beyond traceable fault. This leads us on to the final question of how AI can be orientated towards the public interest.

○ *Artificial intelligence should be developed for the common good and benefit of humanity*

There is very little in what the Government has said to suggest that it has any broader conception of the 'common good' or the public interest than economic growth. The recurring theme is of a 'pro-innovation legal regulatory regime', in which the regulation in question is largely self-regulation and the role of the state is fairly limited to providing certain 'safeguards' or minimal accounts of the public interest pertaining to 'real risk'.

Rather than setting up a dedicated regulatory public body to help govern the impact of AI,[53] the Government instead introduced several non-statutory bodies and corollary initiatives to facilitate the new 'agile pro-innovation approach' to AI governance such as the Centre for Data Ethics and Innovation (CDEI). The CDEI is the principal advisory body to the Government on AI policy, comprised of academics and industry figures. It has no regulatory power and its advice is limited to responding to questions set by Government. For example, in their review, The Review into Bias in Algorithmic Decision-Making. (CDEI, 2020), the Centre was limited to focusing on the public sector and consequently made the following three[54] recommendations:

---

[53] Matt Hancock, Minister for Digital, Culture, Media and Sport, January–July 2018, explained that the Centre 'will not be a regulatory body' but that it will 'ensure that the adoption of AI is accompanied, and in some cases led, by a body similarly set up not just with technical experts who know what can be done but with ethicists who understand what should be done to make sure that the gap between those two questions is not omitted' (House of Lords Select Committee, 2018: 354). Despite the Human Fertilisation & Embryology Authority (HFEA) being a regulatory public body, Hancock went on to cite it as the example to follow in ensuring that 'society moves at the same pace as the technology'. I shall discuss this issue in the following chapters.

[54] Summary available at: www.gov.uk/government/publications/cdei-publishes-review-into-bias-in-algorithmic-decision-making.

(1) Mandatory transparency obligation on all public sector organisations using algorithms that have an impact on significant decisions affecting individuals;

(2) Organisations should be actively using data to identify and mitigate bias;

(3) Government should issue guidance that clarifies the application of the Equality Act to algorithmic decision-making.

My sense is that these recommendations, as important as they are, are not quite a match for the title of the review The Review into Bias in Algorithmic Decision-Making. and are, instead, reflected largely in the guidance that already exists. However, to its credit, the CDEI has attempted to engage the general public in a far more comprehensive way than many of the other bodies of Government through its 'tracker survey', which monitors how public attitudes to the use of data and data-driven technologies change over time. Its methods include polling with a nationally representative sample of 4,000 individuals (as well as a further 200 without access to the internet) to collect citizens' views. While a wholly valuable exercise in generating new topics of broader discussion around AI, the problem with polling as a way of engaging the public is that it only tells you what the public already thinks about a narrow set of questions posed. Even when the focus groups are methodologically incorporated, the findings are limited. For example, in preparation for drafting the 2023 White Paper 'A pro-innovation approach to regulating AI' (HM Government, 2023a), one of the primary objectives of the CDEI report (2022) on AI governance, entitled 'BritainThinks', was to discover '[w]hat the public expect the government to be doing to keep them safe in regards to AI governance'. The report concluded with five findings:

1) The public continue to have limited awareness of AI ...

2) The benefits of AI are broadly seen to outweigh the risks ...

3) Participants' views on governance around transparency and account-ability are tied directly to the perceived risk of AI's use in any given context ...

4) Low familiarity with more complex AI applications makes it difficult for participants to specify what governance they expect ...

5) The public therefore expect AI Governance to work with these principles but also to develop ahead of detailed public understanding
and expectation.

These results are plausible but, as I shall explore in Chapter 4, I am not
convinced by the approach adopted by the CDEI. Is it any wonder that the
people who do not understand the risks of AI do not push for firmer regulation? With this sort of method, there is little room for engagement with
alternative experiences, debate or what Young called 'self-transcendence'
(1997: 66) – the development of a willingness to be open to a different way of
conceiving of and solving collective problems altogether. These findings, to my
mind, do not suggest that the best way forward is for the government to avoid
legislation that could 'stifle' innovation as the CDEI and government suggest.

While offering some general non-statutory principles such as security,
explainability, fairness, accountability and redress, and suggesting that
these ought to be enshrined in a regulatory duty at some point, the
2023 White Paper says little about how to implement these, even in terms
of establishing liability: 'it is not yet clear how responsibility and liability
for demonstrating compliance with the AI regulatory principles will be or
should ideally be, allocated to existing supply chain actors within the AI
lifecycle' (Office for Artificial Intelligence, 2023: 55).

In addition to the CDEI, other non-regulatory advisory bodies to the
Government include the Office for Artificial Intelligence (responsible for
implementing the 2021 National AI Strategy (HM Government, 2022d)),
and the AI Council, which is tasked with providing leadership on the adoption of AI across the UK economy. Most recently, the UK introduced the
Frontier AI Taskforce and the AI Safety Institute following the UK's recent AI
Safety Summit which saw Rishi Sunak interview the likes of Elon Musk. These
initiatives, whilst welcome and necessary, are all populated by field and
industry experts and are focused very squarely on liability-based risks such
as cyber-security and the production and distribution of deepfakes.

It is difficult to see how these bodies have either the remit or powers
to consider the structural dynamics of AI.[55] Indeed, in a recent

---

[55] www.gov.uk/government/news/leading-experts-appointed-to-ai-council-to-supercharge-
the-uks-artificial-intelligence-sector.

2023 YouGov Poll, 66 per cent of respondents reported that they were not confident that technology companies were developing AI responsibly, and 68 per cent reported they were not confident that the UK government was effectively regulating the development and use of AI.[56]

Overall, in its quest to be a 'global AI superpower', the approach of the UK to AI governance is weak on liability-based solutions, and has no capability to address structural dynamics that operate beyond traceable liability, in the wider public interest.

POLITICS AND AI

The leading futurist Amy Webb, whose views framed the very beginning of this book, sets out two futures for the new AI age (Webb, 2023). The first future is the sort of optimistic account of AI that we see in the domestic and foreign policy initiatives of states and international organisations, and the glossy materials of management consultancies and corporates – a future where AI development is grounded in the public interest, with transparency requirements built into AI system design, where humans are in the loop to provide oversight,[57] individuals have control over their personal data, and where AI is used to support human labour rather than replace it. The second future is 'catastrophic', where power is concentrated in a handful of AI companies who 'aggressively curate' user 'needs' according to their own interests with very negative all-round effects for the rest of us humans.

Webb gives the optimistic future a 20 per cent chance (Webb, 2023: 35.48).

This is because AI is no longer trained and managed on given datasets like it used to be, nor does it predominantly operate according to user search prompts. Rather, new forms of AI collect all possible scrapable data on a continual basis, including new original data which is produced

---

[56] See the following link for more details on the YouGov survey (for which 2,011 adults in Great Britain surveyed 5–9 May 2023): https://docs.cdn.yougov.com/oubcmy6k8x/Internal_AI_230509.pdf.

[57] See, e.g., Li et al. (2023) on reinforcement learning from human feedback (RLHF). This is where an algorithm learns from human feedback rather than relying on an engineered function.

by AI itself. This results in cumulatively unfathomable quantities that are so vast they can only be processed on a scale that a handful of tech companies can handle with cloud technology which in turn clusters capabilities, and therefore new AI products, into a very small number of companies. It is now likely that the level of AI surveillance through personal data collection will become absolute (every background detail of a life will be collected surreptitiously – who visited, what music was on in the background, how long a door was open for, etc.), that eventually, with no user control, our daily functions will become heavily choreographed by what a given algorithm predicts we ought to be doing or seeing or choosing, rather than what we might actually like to do differently such as see something new or choose an alternative. As AI becomes an increasingly standard feature of everyday product design, we will have less opportunity to avoid its generalised collection of our data. A fundamental problem with AI is that it tends to make a leap from correlation to prediction, and is unable to accommodate the uncertainty of causation or the possibilities of a desired change (think back to the Amazon HR recruitment algorithm, for example). The net result is an alarmingly powerful choreographer of the behaviours of large populations, a reduction of creative human plurality to a curated set of ground truths and a consolidating cycle of power concentration in a small number of tech companies that is inherently oriented towards its own interests, some of which inspire recklessness. Allen argues for example that it will be companies such as OpenAI (which called its own GPT2 product 'dangerous' and yet released it anyway) that will be calling the shots in the future (Allen, 2023: np). AI scientist Stuart Russell lamented in his 2021 Reith Lecture that the 'success of my field poses a risk to my own species'. One thing is clear, the 'success' of Russell's field would be the failure of mine – politics.

As we have seen, there is plenty of scope for AI to contribute to injustice, both liability based (in the form of algorithms that compound traceable bias and discrimination) and also structural injustice (in the form of mass collective outcomes manifest in numerous debilitating power dynamics and extreme material inequities within the context of a relational social matrix). Societies' increasing reliance on AI and humans' capacity to develop an ever more thoughtless relationship with

AI is a serious problem for politics. As Runciman argues, '[p]olitics needs to regain a measure of control over these machines and over the people who currently control them. Otherwise, there is a danger that, instead of using the machine to help solve our problems, we limit ourselves to the kind of problems that can be solved by machines' (2018: 126). Benjamin reminds us that we are the 'pattern makers' and in this realisation, that our habits and behaviours (including in large part our structural actions) generate the data that powers AI, we can better understand our relationship to the structural dynamics of AI.

The UK, like all other states, has adopted a liability-based approach to the governance of AI which is, I have suggested, ineffective for addressing the increasingly important structural dynamics of AI, which require a more speculative approach. Thinking back to the question that framed Chapter 1 – *Where nobody is liable, who is responsible?* – we find little answer to that question in the current approaches to AI governance.[58]

While numerous background social processes and relations operate behind a range of structural injustices, AI, as an increasing generic transitional technology, is bound to be heavily imbricated across these structural matrices. It is not sufficient to rely on somewhat narrow philosophical, or indeed, at the other end of the methodological spectrum, social scientific requirements that need a clear causal line between agent and outcome to be established for political action to ensue. My view is that Young's account of structural injustice alludes to the reality that politics is far messier than that, and yet, nevertheless, we are required to generate a coordinated response.

Failing to govern AI might just turn out to be humanity's greatest political mistake. In a world where companies are tracked internationally and tightly regulated nationally on new drug releases or nuclear energy production, tech companies are at liberty to release, often unpolished, transformative AI, such as ChatGPT. Indeed, as Barrat notes, tech companies' 'only regulation is their quarterly profit report. They're not going to take it upon themselves to regulate the industry. We the people are

---

[58] This is a different consideration from that which Floridi (2016) raises about 'faultless responsibility' ending in criminal behaviour (this might be akin, e.g., to indirect discrimination) which will be grounded in the traceable actions of a liable agent.

going to have to do it' (2023: np). How we do it, exactly, is a difficult political question to answer, and I return to this major theme in Chapter 4.[59]

First, however, I turn to another extraordinary technology, Repro-tech, where there is a different story to tell about structural injustice and governance.

---

[59] It is worth noting that even Zuckerberg surmised, 'I actually am not sure we shouldn't be regulated. I think in general technology is an increasingly important trend in the world. I think the question is more what is the *right* regulation rather than "yes or no should we be regulated?"' (Patel, 2018).

CHAPTER 3

Repro-tech and the Genetic Supermarket

[A] 'genetic supermarket', meeting the individual specifications (within certain moral limits) of prospective parents ... This supermarket system has the great virtue that it involves no centralized decision fixing the future of human type(s).

Nozick (1974: 134)

I predict that by 2050, if I am a young woman and ... I want a child, I'll do it by IVF ... Women will in fact eventually decide to be fertilized by IVF methods for the simple reason that concurrent with the improvements in assisted reproductive techniques, enormous advances have occurred in the area of genomics ... the ultimate factor.

Djerassi (2014a: 74)

BOTH OF THESE PREDICTIONS, ONE BY THE LIBERTARIAN philosopher Robert Nozick and the other by the leading reproductive technology scientist responsible for the development of the contraceptive pill, Carl Djerassi, would have seemed fantastical only a few years ago. However, today the idea that fertile women could undergo elective 'routine' oocyte cryopreservation, otherwise known as 'proactive' egg freezing, and employ *in vitro* genetic technologies to alter characteristics of their future child, is wholly conceivable today for those who have the financial means. In this chapter, I want to explore some of the potential structural dynamics of reproductive and genetic technologies – 'Repro-tech' – and in particular, the 'gold standard' governance approach of the UK, which boasts the first and oldest Repro-tech regulator in the world. Despite some vitally important elements to the UK's approach which I shall build on in Chapter 4, I hold that, akin to the context of AI, a governance mechanism grounded in liability is far from sufficient for addressing the wider structural dynamics of such a potentially transformative technology as Repro-tech.

REPRO-TECH

Certainly, there is no doubt that reproductive technologies have had, and will continue to have, a profoundly positive impact on the lives of many people seeking to become parents. Some of these have fertility conditions, some use the technology to avoid passing on inherited genetic conditions, some wish to use Repro-tech to have children with a same-sex partner or on their own, or alternatively, use these technologies as an insurance policy for the future[1] (see, e.g., Baldwin (2017); Baldwin et al. (2018); Golombok (2015); Waldby (2015)). Indeed, Inhorn's most recent work, *Motherhood on Ice* (2023), based on interviews with 150 women in the United States who have frozen their eggs, investigates the complexities that surrounded these women's decision to do so. Of those 150, 114 women froze their eggs for 'non-medical' reasons.[2] Inhorn discovered that since the American Society for Reproductive Medicine had permitted egg freezing for 'non-medical' reasons in 2012, more than 36,000 women had frozen their eggs. The majority of these women were in their 30s and were motivated to freeze their eggs due to what she describes as 'partnership problems'. Indeed, 82 per cent of these women were single at the point of freezing.

Here, however, I want to focus on the structural dynamics of proactive egg freezing which serves as 'fertility insurance', and its relationship to emerging pre-implantation genetic technologies (PGTs).

Now commonly provided as an employment benefit by some of the world's most famous companies, including Facebook, Apple, Google, Unilever, Deloitte, Uber, LinkedIn, Intel, eBay, Yahoo, Netflix, Salesforce, Spotify, Time Warner, Snapchat and Goldman Sachs,

---

[1] In the last credible assessment, over 9 million children were estimated to have been born through Repro-tech across the world in 2020 (Kuhnt and Passet-Wittig, 2022).

[2] The language of 'non-medical' is in some sense unsatisfactory because naturally declining fertility is of course a medical issue, but here Inhorn is using a common label for such a procedure.

Repro-tech is fast becoming a viable option for those employees making reproductive choices. The following advert by Repro-tech company Spring Fertility exemplifies the sort of promotional material that many young women will be familiar with:

> The benefit of egg freezing is time – the time to pursue your dream career, to travel, to meet the right partner – all with the peace of mind that you'll be able to start your family when you are ready.[3]

This sort of 'fertility insurance' is aimed at women who can secure the means to try to buy time on their biological clocks. As I shall discuss, the primary drivers of this emergent fertility insurance industry are twofold. The first is of little surprise – the prevalence of low replacement birth rates in many wealthy economies where women tend to have (fewer) children later in life. The second is the increasing attraction of PGTs. These new technologies began with the opportunity to genetically screen for a range of conditions but, as Nozick predicted, the promise of gene trait selection is continuously growing in the public imaginary. Before exploring these PGTs in a little more detail, I shall give a brief explanation of what egg or embryo freezing technology entails.

Women are born with all of their 'potential eggs' (follicles), but by the time the average woman is in her early to mid-30s, these numbers will have declined dramatically: she will have approximately 12 per cent remaining, which will reduce to about 3 per cent by her early 40s and continue to decline until the menopause at around 50 years of age (see, e.g., Park et al., 2021). Women who decide to freeze their eggs for future use, first undergo a range of medical tests, followed by two to four weeks of hormone treatment to stimulate the ovaries (superovulation) to produce multiple follicles/eggs.[4] This process increases the production of follicle/eggs to as many as 40 per cycle via superovulation (compared to the typical production of only 1 follicle/egg in a regular menstrual

---

[3] See https://springfertility.com/egg-freezing/.

[4] Superovulation is not without its risks. The collection of high numbers of eggs per cycle can result in ovarian hyperstimulation syndrome, potentially affecting ovaries and major organs such as kidneys, lungs and liver (see, e.g., Aljawoan et al., 2012), as well as ovarian cancer (van Leeuwen et al., 2012).

cycle). Once the follicles are sufficiently large, the eggs are collected.[5] The number of eggs retrieved can vary significantly, but often the recommended minimum number for freezing is between 7 and 14 eggs. The harvested eggs are then dehydrated, treated with cryoprotectant ('anti-freeze'), frozen (through a flash-freeze process of vitrification)[6] and stored in liquid nitrogen for up to fifty-five years (egg freezing). When a woman who has frozen her eggs wishes to attempt to become pregnant, her eggs are defrosted and then fertilised by the IVF intra-cytoplasmic sperm injection method (ICSI), whereby one sperm is directly injected into a thawed egg to heighten the chances of successful fertilisation. Multiple embryos are created using this technique, and having matured for four to five days, one or sometimes two are chosen for implantation into the patient's uterus in the hope of pregnancy (see, e.g., Estudillo et al., 2021). Alternatively, the harvested eggs can be combined with sperm to create an embryo which is then frozen, to be defrosted and implanted later.[7] While there is no data available on the success rates of frozen eggs to birth, the average IVF birth rate using frozen embryo transfers stands at 27 per cent, with dramatically different

---

[5] Retrieval is conducted using a long needle passed into the ovary through which the eggs are collected by suction (transvaginal ultrasound aspiration).

[6] The human egg is high in water content and consequently prone to ice crystallization, which can cause significant cryodamage to genetic material in the freeze–thaw process (see, e.g., Rodriguez-Wallberg and Oktay, 2012). Vitrification, designed to reduce cryodamage, is a technique that involves a much faster flash-freeze process than the previous slow-cooling methods. The first child to be born in the UK using the vitrification technique on a patient's own eggs (frozen for five months), Olivia Bate, was born in 2010 at the Midland Fertility Clinic, Tamworth (Midland Fertility Clinic, 2012).

[7] Risks to women undergoing the treatment include ovarian hyperstimulation syndrome (a reaction to fertility drugs which can cause stomach pain, nausea and faintness) in the context of egg collection, a higher likelihood of multiple-birth pregnancy (and the associated risk of neonatal death, late miscarriage, high blood pressure and gestational diabetes), as well as ectopic pregnancies and birth defects for the child. Women who become pregnant over forty years of age face significantly higher risks of thrombosis, hypertension, gestational diabetes and pre-eclampsia. (See HFEA 2023 'Risk of Fertility Treatment' website. Available at: www.hfea.gov.uk/treatments/explore-all-treatments/risks-of-fertility-treatment/). Although technologically spectacular, the process of egg freezing and IVF/ICSI can present women with significant health risks, and there remain considerable gaps in knowledge on the long-term health risks of humans born via the vitrification technique, which is a relatively new technique: see Footnote 6.

success rates according to the age of the patient at the time of freezing (HFEA, 2023a).

DRIVERS OF PROACTIVE EGG FREEZING

> Friends Freeze Together! Announcing our 2022 egg freezing special from our industry-leading fertility clinic ... 3 friends freeze, each receive 30% off: 2 friends freeze, each receive 25% off.
>
> Southern California Fertility Clinic[8]

The backdrop to this new fertility insurance market aimed at fertile women with sufficient financial resources is predicated on what Djerassi (2014b: 73) called the 'mañana [tomorrow] generation' – a generation defined by delayed reproduction resulting in 'geriatric states' (5). These are states where the total fertility rate (TFR) is less than 2.1 – the rate required for population replacement. The TFR for the United States and China is 1.7 (World Population Review, 2023); the collective TFR for Europe is 1.53, and not a single EU country has a TFR above 1.90 (Eurostat, 2023).[9] Indeed, the number of births across the EU has dramatically declined over the past forty years. The key factor in this downward trend in many countries is that women are having children later in life and, consequently, fewer of them. In 2022, the mean age of women at childbirth in the EU was 29.4 years, and 30.7 years in the UK where the TFR is currently 1.7 (World Population Review, 2023). The reasons for this trend are familiar. Over the past thirty years women have become as likely as men to stay in higher education, accumulate greater debt,[10] and are economically dependent on employment. By way of

---

[8] Advert by Southern California Fertility Clinic – See advert: https://www.facebook.com/photo?fbid=10160496834302125&set=a.10150251832962125&locale=hi_IN.

[9] Fertility rates under 2 are not universal. In Sub-Saharan Africa, e.g., the collective TFR stands at 4.6 – the highest in the world, despite a substantial decrease from 6.3 over the past 30 years (UN, 2021a). Research by the African Network and Registry for ART shows that access to, and usage of, Repro-tech remains relatively low across Africa (Dyer et al., 2020).

[10] The average personal debt in the UK in March 2022 was £33,410 (https://moneynerd.co.uk/average-personal-debt-uk-2022/) and the average household debt stood at £63,582 (https://themoneycharity.org.uk/money-statistics/march-2022/).

example, Djerassi pointed to the UK female graduate population as a typical 'mañana generation' where the more normalised late-motherhood becomes, the more attractive (and rational) fertility insurance will become for those who can afford it (Djerassi, 2014b: 73).[11] As van de Wiel (2020a) argues, '[b]eyond a treatment for current experiences of infertility, IVF is increasingly oriented towards the pre-emptive and proactive treatment of future infertility' (306).

In 2019, Goldman Sachs introduced its 'Pathway to Parenthood' programme, which provides bursaries up to $40,000 for costs relating to adoption, egg freezing, egg donation, IVF and surrogacy. In a similar vein, Richard Branson (founder of Virgin Group, which also has a similar policy) celebrated the corporate turn towards Repro-tech, '*it's women's choice*. If they want to carry on working, they can carry on working. If they have not met the man of their dreams at 35, 36, 37, 38 – freeze the eggs!' (Branson, 2015; emphasis added).

While individual employers are keen to support their employees and remain competitive in their retention packages with peer institutions, the sort of structural perspective discussed in Chapter 1 can help us to think fruitfully about some potentially troubling aspects of the promotion of proactive egg freezing as a means of exercising individual responsibility to better managing career prospects. Proactive egg freezing often tends to be viewed as simply expanding individual women's reproductive choice. However, these sorts of arguments are too narrow in that the framing of proactive egg freezing as a rational choice made by the responsible individual is over-emphasised and the collective social aspects of technology use are underplayed.

Thinking back to the discussion of structural injustice in Chapter 1, the propensity to see people's life decisions (material and social resources as well as their future prospects), too much in terms of liability is likely to lead to an insufficient account of, and response to, social structural processes. As Haslanger describes, these sorts of processes impose constraints on our action, they render us into social positions within a set of relations that in turn constitute further social structural processes (2016: 125). This is what she calls the 'choice architecture'

---

[11] Also see Bendix (2014).

which impacts on the options and choices of each individual in the structural matrix (2016: 127). In this way, we are all contributing to the background conditions of each other's prospects. With respect to the choice architecture of our reproductive choices, there is tension with economic production familiar to many feminist writers, which is worth repeating until solutions are found. For example, Fraser (2021) describes how under the neoliberal conditions of contemporary capitalism, social welfare provisions are systematically diminished and care is externalised to the private spheres of individuals, families and communities while their wages stagnate and their hours of work increase. The result is the increased incapacity of individuals, families and communities to perform caring duties in a context where debt is one of the defining features of our 'gig-economy' era (2021: 159).

For Fraser, one of the most disappointing political aspects of contemporary democratic politics is the development of what she labels 'progressive neoliberalism' (2021: 160): a redefinition of emancipation in market terms where objectives such as diversity, gender equality and meritocratic pay structures become markers of a successful labour market while increasing externalisation of social reproduction (with its own racialised, gendered and socio-economic dynamics, not least in global care chains) continues apace, largely under the contemporary political radar. By way of example, Fraser describes the neo-liberal imaginary of gender equality as one which features women who are 'deserving of equal opportunities to realise their talents, including – perhaps especially – in the sphere of production. Reproduction, by contrast, appears as a backward residue, an obstacle to advancement that must be sloughed off, one way or another, *en route* to liberation' (2021: 161). Such an imaginary has produced a society 'obsessed with technology' as having all the answers to society's problems, and here she invokes the example of proactive egg freezing being provided by a growing number of employing institutions eager to attract and retain staff: 'wait and have your kids in your 40s, 50s, or even 60s; devote your high-energy, productive years to us' (2021: 162).[12]

---

[12] Fraser also discusses breast pumps in a similar way: 'In a context of severe time poverty, double-cup, hands-free pumps are considered the most desirable, as they permit one to express milk from both breasts at once while driving to work on the freeway' (2021: 162).

Let us imagine a scenario linked to the rise of these private provisions of proactive egg freezing. Bella and Stella are employees of the same organisation – Company Z. They are equally intelligent, talented and experienced in their work, and both earn a relatively good salary. Both have sizeable student debt, having invested in education to progress up the career and salary ladder and are keen to work their way up to a higher wage to pay off those debts and get onto the property ladder. Both are in their early 30s and neither have fertility challenges beyond those associated with age. Both are in long-term relationships; neither have children but would like to in the future. On the face of it, both Bella and Stella have very successful and privileged lives. Company Z has a fertility insurance policy whereby the considerable costs of proactive egg freezing and storage are covered for those who choose to take out fertility insurance. Bella is incentivised to take up the fertility insurance and undergoes egg freezing with a view to focusing on her career for the next decade. Stella is more hesitant to undergo the necessary hormone treatment and decides to wait. Company Z is following other leading companies in adding proactive egg freezing to its array of health policy offerings. Egg freezing is prohibitively expensive for many women, and the option to take advantage of free fertility insurance is welcomed by employees, whether or not they take it up themselves. Bella makes the right decision for Bella. She does not want to risk losing an opportunity that might give her the edge on applying for a higher paid role that comes up, and is relieved that proactive egg freezing could give her a longer timeframe in which to manage her fertility and economic objectives. Stella, on the other hand, is weary of the considerable health implications of proactive egg freezing and decides with her partner to try for a baby in the coming year, with a view to returning to work as soon as she can. Stella makes the right decision for Stella. This story could end in several ways, of course, but let us consider some possible consequences of Bella's and Stella's choices.

The rational calculus (whether legal or not) to favour the worker who is 'unincumbered by biology' is already apparent in many a study of the gendered workplace, both historical and contemporary (Browne, 2006; Haslanger, 2016; Gatrell et al., 2022). However, the perceived fault line that used to run through a workplace, based on a simple assumption about gender, is now far more complicated by Repro-tech. The idea that

'Smart women freeze' (Money-Coutts, 2021: np) lends itself to the narrative that those who 'take out fertility insurance' are the better fit for the high-flying jobs that demand dogged commitment and a seamless management of the complications of personal life and 'leisure time lifestyle'. Perhaps women who make the same choices about Repro-tech as Bella begin to look more committed to their work than those who align with Stella's choices, and an unspoken division of opportunity and material resources begins to grow. How would a liability-based approach account for this? One answer is to say that the decisions taken are merely evidence of revealed preferences – either as Haslanger (2016: 122) caricatures; 'women prefer motherhood over committing to the most demanding careers, so they choose to forego the opportunities offered to maximise career progression' or as Tarasoff et al. (2014) describe, the promotion of social egg freezing creates 'a moral imperative to engage in social egg freezing ("just in case"). That is, if we are women who have the option to freeze our eggs, then we should do so, and any negative consequences arising from our failure to control the future through our decision not to freeze our eggs are our responsibility and fault alone' (2014: 239).[13] Those who have been 'responsible', however, in freezing their eggs will reap the benefits of sufficient financial security and status, which in turn will provide the right window in which to thaw eggs and start a family. This view is in line with Goold and Savulescu, who have argued that: 'There are reasons to think it is actually better for women to have children *later* in life. Many women who have children when they are older will have higher incomes' (2009: 54). Another liability-based answer might be that institutions and the state should provide much better care policies to enable parents to have children at any point in their lives, not just when they have got past the pinnacle of their careers. While many companies and national public policy programmes have begun to better recognise primary care roles as an essential element of social reproduction (to put it in Fraser's terms), the lack of provision of comprehensive childcare facilities and financial support for those who

---

[13] Also see Phillips (2016). Another relevant point here is made by Sherwin (1987) that the promotion of reproductive technologies serves to remind career women that however satisfying their careers are, they should nevertheless want to have children eventually.

care at home and in their communities are intrinsically woven into the background conditions of millions of peoples' experience of structural constraints. I have written elsewhere on the likely effects of comprehensive child and elder care schemes in conjunction with well-paid care leave schemes that enable all parents and carers to take up a primary-care role, whether or not they have full careers or are full-time carers (Browne, 2013c).[14] Certainly, effective care policies operate as a genuine facilitator to those who want to become parents in the *present* rather than delaying childbearing until their supposedly 'wealthier future', as Goold and Savulescu advocate. As Fraser (2021: 163) argues, the negative effects of these sorts of policies are not isolated. There is, in many economies, a distinct lack of sufficient 'social arrangements that could enable people of every class, gender, sexuality, and color to combine social-reproductive activities with safe, interesting and well-remunerated work'. These answers provide some sort of traceable account to individuals, groups and institutions, but what might be the structural dynamics of proactive egg freezing?

With increasing provision of fertility insurance policies as a marker of prestigious employment, it is possible that the more traditional solutions of comprehensive state subsidised childcare and paid parental leave (which are yet to be achieved, as many feminist scholars have pointed out over many years) would soon be 'crowded out' as a viable nationwide programme of policy reforms. The argument here is not just to point to a failure of particular policies (as Young argued, this can be captured by the liability-based approach) but rather to imagine a future where such

---

[14] To return to Goldman Sachs again, twenty weeks' paid parental leave is now offered to all employed parents. This is a vastly improved policy on the national provision in the United States where there is in fact no federal provision for paid parental leave. In the UK there is a paid parental leave provision whereby mothers are able to claim six weeks of 90 per cent pay after which they can share up to thirty-seven weeks paid leave with their partner at the rate of £172.48 a week. To put this into context, the average weekly earning in the UK is £596 per week (UK Government Maternity pay and leave2024)) so it is not surprising then that take-up of the shared parental element is very low. Most company-specific schemes are not only inevitably limited to their employees but also too short, unsupported by affordable childcare alternatives, insufficiently paid and too restricted to the mother, thereby locking in a gender split even when parents wish to take on more equal primary parenting roles.

policies begin to lose their logical appeal, where it seems like a bad decision not to take out fertility insurance like any other insurance, even when resources are tight and coupled with a diminishing set of workable care services. As Haslanger points out, '[t]he focus on individuals ... reinforces fictional conceptions of autonomy and self-determination that prevents us from taking up responsibility for our social milieu' (2015: 10).

Gürtin and Tiemann (2021) document how fertility insurance is predominantly provided by the private sector, and as van de Wiel (2020a) argues, the marketisation of fertility dovetails neatly with neoliberal objectives of human capital maximation in an economy that creates heavy incentives to put motherhood on hold. After Facebook's proactive egg freezing policy announcement, fertility platform EggBanxx[15] adopted the slogan 'Lean in. But freeze eggs first!', referring specifically to Facebook COO, Sheryl Sandberg's book (2013), *Lean In: Women, Work and the Will to Lead*, which argues that women ought to be much more assertive at work in order to succeed professionally. Gina Bartasi, former CEO of EggBanxx and founder of Kindbody that provides corporates with fertility insurance services and worth $1.8 billion (CNBC, 2023), is noted for normalising fertility insurance: 'Lean In and live your career dreams – But Freeze First. Do not find yourself with regrets. Make options for your future self, now'. As van de Wiel (2020b: np) argues the sort of framing of fertility that relies on the logic, '"you'll never be more fertile than today"... invokes a spectre of loss and scarcity that drives the popularization of new forms of biotechnological dependency, which is presented as "safe" and "work[ing]" in spite of [egg freezing's] limited success rates and potential side effects'. Such provisions are inevitably uneven, incomplete and unlikely to engage with underlying structural issues. As employer-provided reproductive insurance increases, the expectation on women to freeze their eggs at their employer's expense is set to increase. Given that the cost of proactive egg freezing, storage, thawing, fertilising and implantation is around £10–15 k[16] (set against a backdrop of increasing student debt, increasing

---

[15] See CNBC interview available at: www.cnbc.com/video/2023/05/10/kindbody-founder-gina-bartasi-on-democratizing-fertility-treatment-care.html.

[16] For the UK, see, e.g., current prices at the London Women's Clinic (available at: www.londonwomensclinic.com/london/treatment_costs).

cost of living and stagnant wages), one can imagine that it might even appear irrational for many women not to take up their employer's offer to freeze eggs when others around are taking 'responsible' measures. These pressures create an example of Haslanger's 'choice architecture' which in turn structures 'the possibility space for agency' (Haslanger, 2016: 127). The structural perspective thus provides an interpretation of the positioning of the individual in the 'space for agency', which cannot be transformed by the intentions or actions of the individual alone.

'THE ULTIMATE FACTOR': GENETIC TECHNOLOGIES

Beyond the potential for wealthy populations to increase the uptake of proactive egg freezing as fertility insurance is another radical Repro-tech development, which Djerassi thought would be the ultimate factor in future reproductive habits.

The prediction that women will 'actively choose social egg freezing as insurance, providing them the freedom, in the light of professional decisions' (Djerassi, 2014a) is built on the perception that using such technologies carry the added benefits of PGTs. As the quantity of a woman's eggs decline, so the genetic 'quality' of eggs deteriorates with age, in the sense that the risk of particular genetic conditions in children born from older eggs is significantly higher with oocyte ageing (see, e.g., Loane et al. (2013)). The conventional method for genetic testing (frequently routine for older mothers) is the amniocentesis procedure which is performed between the eleventh and twentieth week of pregnancy and which comes with a 1 per cent risk of miscarriage.[17] Franklin refers to the procedure as 'high-tech reproductive roulette' (2013: 7) and if a concerning genetic condition is detected via amniocentesis then for some prospective parents this will result in the decision to terminate the pregnancy. However, if a woman were to choose the alternative IVF route to pregnancy, PGTs can be used to test for particular genetic

---

[17] The procedure consists of inserting a syringe into the amniotic sac that surrounds the foetus. This risk figure of 1 per cent is for the UK; see the UK National Health Service guide for more details at www.nhs.uk/conditions/Amniocentesis/Pages/Introduction .aspx.

conditions in embryos *before* they are selected for transference into the womb, thereby significantly reducing elective terminations or the number of children being born with debilitating or life-threatening genetic conditions. In the UK, for example, there are over 600 genetic conditions that can be tested for by PGTs (such as muscular dystrophy, haemophilia, Down's syndrome and cystic fibrosis (HFEA, 2023b)). By producing multiple embryos *in vitro*, testing with PGTs for certain conditions and then only selecting embryos without those conditions for implantation in the prospective mother's womb, debilitating and life-threatening genetic conditions could be potentially eliminated (Metzl, 2019) if sufficient access to these technologies was provided globally. Savulescu (2020) calls this use of PGTs 'liberal eugenics'.[18] Djerassi thought that as women increasingly opted for proactive egg freezing, it would inevitably follow that genetic screening would become a standard part of the process. This, he predicted, would open the door to large-scale gene editing.

There are broadly two forms of gene editing – somatic and heritable germline. Somatic editing is where the DNA of an individual is edited to address a particular single-gene condition. In 2015, UK doctors at Great Ormond Street Hospital in London used a somatic gene editing technique (Transcription Activator-Like Effector Nucleases – TALENs) whereby donor immune cells are genetically engineered to target cancerous cells while ignoring the drug, alemtuzumab, used to suppress the immune system, so that the new cells are not rejected when administered to a patient. As a world-first procedure, these genetically engineered immune cells were injected into a one-year-old child named Layla Richards, who was suffering from a particularly aggressive form of leukaemia. It is reported, to date, that Layla remains in remission. As one of the responsible doctors, Paul Veys, described the procedure as 'almost a miracle' (quoted in Sample, 2015). In heritable germline editing, oocytes or early-stage embryos are genetically modified, which results not only in a baby being born with all their cells modified but any related future offspring will also inherit these DNA. Controversially in 2018, He Jiankui, a scientist at Southern University of Science and Technology of China in Shenzhen,

[18] See also Savulescu (2014).

pronounced on YouTube that using a gene-edited technique called Clustered Regularly Interspaced Short Palindromic Repeats (CRISPR), he had created the first genetically edited babies. He was subsequently imprisoned by the Chinese government for illegal medical practices.[19] Inevitably, however, heritable gene editing is on its way and, as Djerassi predicted, there is a growing consensus that the reproductive habits of the future will be fundamentally influenced by the ability to select for certain traits through PGTs, at least for those who can afford it. Why might this be a structural consideration?

HUMAN ENHANCEMENT AND THE GENETIC SUPERMARKET

Renegade scientists and totalitarian loonies are not the folks most likely to abuse genetic engineering. You and I are – not because we are bad but because we want to do good . . . The most likely way for eugenics to enter into our lives is through the front door.

Caplan interviewed by Metzl (2019: 175–176)

Bostrom surmises that as the genetic supermarket 'becomes more common, particularly among social elites, there might be a cultural shift toward parenting norms that present the use of selection as the thing that responsible enlightened couples do', adding that, '[m]any of the initially reluctant might join the bandwagon in order to have a child that is not at a disadvantage relative to the enhanced children of their friends and colleagues' (2014: 39).[20]

---

[19] See this YouTube announcement posted by He Jiankui, a scientist at Southern University of Science and Technology of China in Shenzhen in 2018: https://youtu.be/th0vnOmFltc. Here it is claimed that He Jiankui had performed the CRISPR process on two *in vitro* embryos to make them HIV-resistant, having fertilised eggs from a prospective mother, Grace, with the sperm of her partner Mark through IVF. He Jiankui explained that Mark carried the HIV virus. These genetically modified embryos were then implanted into Grace's womb and subsequently were born as twins Lulu and Nana. Heritable germline editing is a highly controversial and embryonic technology that is yet to be properly tested in human trials. Certainly, the health consequences of germline editing are an unknown. His pronouncement sent shock waves through the global scientific community, and according to Devlin (2023), he was sacked from his university position and sent to prison for three years by the Chinese government.

[20] See also Shulman and Bostrom (2014).

Savulescu (2020)[21] calls for 'rational evolution' whereby gene editing actively enhances the next generation: 'we could have IQs vastly greater than humans currently have today'. Bostrom suggests that such an approach[22] would be not only positive for individuals and families but for the wider society through substantial increases in human capital and economic growth (Shulman and Bostrom, 2014: 189).[23] Similarly, MacAskill (2022: 145–146) argues that '[a]dvances in biotechnology could provide another pathway to rebooting growth. . . . if human beings were genetically engineered to have greater research abilities, this could compensate for having fewer people overall and thereby sustain techno-logical progress'. Savulescu (2020: np) continues that 'we may no longer be the same species. Genes could be transferred from non-human animals to humans to provide them with bat-sonar, hawk-eye vision, enhanced memory or even radical life extension'.[24] Other possibilities are, for example, 'increased immunity and resistance to disease, toler-ance for adverse environmental conditions (such as that of space), super-abilities or other various factors such as the ability to make vitamins rather than having to consume them' (Gyngell et al., 2019: 516). Bearing these possibilities in mind, Savulescu (2014) argues that genetic modification is a moral obligation and parents should be allowed to access this technology and have children with what he sees as better prospects.

MacAskill (2022) takes this sort of thinking a step further in his belief that genetic technologies will lead to the fusion of artificial general intelligence hosted by synthetically produced life forms. This fantastical sounding prospect is closer to reality than we might think. Take, for example, the Cambridge University team led by Magdalena Zernicka-

---

[21] https://dohadebates.com/video/julian-savulescu-genetic-enhancement-is-a-moral-obligation/.

[22] Some of this sort of thinking has been linked to contemporary racism. Take, e.g., Bostrom's claim that 'IQ correlates negatively with fertility in many modern societies' as reported by Torres (2023).

[23] See also Bostrom and Sandberg (2017).

[24] I cannot help but be reminded here of the well-known politics article that Nagel wrote in 1974 about how humans could never really know what it would be to experience the life of a bat. Presumably, Savulescu might not entirely agree that will always be the case.

Goetz in 2022, who successfully developed a synthetic mouse embryo entirely from stem cells which self-organised to produce such functions as a beating heart and the foundations of a brain (Collins and Garget, 2022). More recently the first human synthetic embryo was successfully grown, again entirely from stem cells *in vitro*. This is a staggering scientific development. Currently these synthetic human embryos develop to the stage of gastrulation, whereby the embryo begins to form the basic features of the body (Bao et al., 2022). This research is intended to help us understand why some embryos fail in pregnancy and also to develop synthetic human organs for transplantation. However, such an extraordinary scientific development is also a landmark to a future where animal life forms, including humans, can be produced synthetically without parents. What would this mean for the ways in which we understand ourselves and relate to each other in future generations? Torres (2023: np) argues that the work of MacAskill and Bostrom are 'longtermists' who believe that the future could see 'a superior new race of posthumans' through a combination of AI and genetic editing which Torres (2022:np) describes as 'eugenics on steroids'.

How accurate these 'positive-trait gene selection' technologies turn out to be is yet to be established. Certainly, there is significant scepticism that genetic dispositions are only part of one's personality in a sea of environmental factors (Dupré, 2021). Nevertheless, there can be little doubt that gene selection will create a lucrative market for those attracted by the allure of creating enhanced children – as Nozick predicted, a 'genetic supermarket'.[25] Once PGT markets are better established, Djerassi argues that, 'IVF will start to become a "normal" non-coital method of having children' (2014b: 74) and therefore, if decided with foresight, the freezing of 'optimal' young eggs years before becoming pregnant will become increasingly popular for those able to pay.

---

[25] The genetic supermarket was a passing idea in Nozick's Anarchy, State and Utopia (1974), expressed in a footnote and set as an example in the future that epitomised his argument for keeping the state out of private decision-making. Nozick argued that prospective parents should be able to choose the genetic features of their children in the context of a free market without any state interference, ensuring that there would be 'no centralised decision fixing of the future human type(s)'. See Gavaghan (2007) for further discussion.

The prospect of a genetic supermarket has inevitably generated much concern. Buchanan et al. (2000) and Habermas (2003) lament the immense potential that gene selection markets would have for heightening inequality and injustice. Similarly, Sandel imagines that a world in which some parents 'became accustomed to specifying the sex and genetic traits of their children, would be a world inhospitable to the unbidden, a gated community writ large' (2007: 86). More recently, Hassan (2020) has argued that we ought to abandon the idea of pursuing an 'enhanced humanity' and instead restrict ourselves to embryo selection as a means of addressing debilitating or fatal genetic conditions, for fear that gene editing will only serve to exacerbate the vast forms of social inequality we already have. She predicts that with gene selection, any advantages that enhanced children have would more likely be attributed to their genetics rather than their probable privileged backgrounds. This observation would no doubt reflect a key structural dynamic in a future populated with genetic supermarkets. One of the central features of the persistent political failure to address increasing economic and social precarity even within wealthy democratic states is the way in which economies rely on human capital metrics. Let us take the UK as an example: currently the UK's human capital stock is valued in the region of £23.8 trillion (ONS, 2022b) which is a putative measure of the total potential future earnings of everyone of working age[26] in the labour market. An extract from the UK Government's website entitled 'What is human capital?' gives a sense of how human capital is assessed:

> it is expected that people with more valuable attributes, such as higher qualification levels, skills and abilities, will earn more in the labour market. Social attributes, personality and health attributes are also reflected in wage rates. For these reasons, human capital in the UK is measured in monetary terms as the total potential future earnings of the working age population. (ONS, 2018)[27]

Human capital dominates modern economic thinking, and if genetic supermarkets were ever to become a reality, then even if that reality only affected a tiny minority of the population, it is very likely that we would

---

[26] Aged sixteen to sixty-six years.　　[27] See also ONS (2022b).

hear pronounced, 'a sizeable gap in human capital is to be expected' on genetic grounds.[28] The particular problem with this sort of thinking is that it justifies a lack of enthusiasm for broad social policies aimed at facilitating greater opportunities for those worse off, on the grounds that genetic enhancement is beyond the social.

Currently, we live in a world where human diversity is a natural occurrence but as Metzl argues, with the power to choose we are likely to be 'hard-pressed to overcome our biases of the present moment' (2019:180). He gives the example of a 2018 study conducted at the University of Pennsylvania which singled out eight genes that have a significant causal impact on skin colour: 'it is likely that parents with different skin tones selecting from among their embryos will in the not-too-distant future be able to choose one that has a lighter or darker complexion ... Choices like these that could reduce human diversity would not only have significant and negative social implications but also potentially expose us to yet unknown-risks' (181). Here there is a semblance of comparison with AI discussed in Chapter 2. There I discussed the tendency for AI usage to result in a reduction of the plurality of human capabilities. Here too is the possibility of genetic convergence around certain selected genes, the consequences of which may take generations to understand. Moreover, a two-tier humanity cleaved by vastly inadequate access to genetic technologies would serve to create ever deeper structural dynamics that the techno-solutionists such as Savulescu, Bostrom and McAskill fail to sufficiently grasp.

Bearing this in mind, and the fact that the policy space relating to reproduction increasingly concedes to the private sector and its interests (Gürtin and Tiemann, 2021), raises the particular question of how structural dynamics affecting the public interest might be addressed

---

[28] Here I paraphrase Nobel Prize winner for Economics Gary Becker, a key figure in the Chicago School of neoliberal economics, who argued that gendered pay gaps could be understood in terms of women's lower human capital being a consequence of women's lower commitment to the labour market than men (1991: 4) irrespective of need, preference or opportunity. I have long been struck by the circularity of Becker's basic argument – why are women paid less? Because they have lower human capital. How do we know this? Because they are paid less. See Browne (2006) for a critical engagement with Becker's human capital theory.

through governance. Whereas I argued in Chapter 2 that the current governance approach to AI was markedly insufficient for both liability-based and structural harms, there are elements of Repro-tech governance within the UK context that offer some inspiration for the better governance of radical and fast-moving transformative technologies in the future (Franklin, 2019a; Harding, 2024; Browne, 2020). However, I shall now reflect on the ways in which the UK's approach, seen by many as the gold standard of technology governance,[29] has been challenged by its commitment to what I regard as an all too limited account of the public interest grounded in liability and economic growth to effectively consider wider structural dynamics.

REPRO-TECH GOVERNANCE: THE EXAMPLE OF THE UK

Repro-tech is part of one of the UK's most lucrative sectors, the Life Sciences, which is worth over £94 billion to the UK economy (HM Government, 2023c). Unlike AI, however, Repro-tech is tightly regulated in the UK by a public body, the Human Fertilisation and Embryology Authority.

Public bodies operating at arm's length from ministers come in different forms,[30] contributing to the processes of governance, devising and delivering on a range of public services and, broadly speaking, working within a strategic framework set by government. In the UK, there are currently 295 such bodies, employing over 300,000 staff, with government funding of more than £220 billion and ranging from NHS England and the Secret Intelligence Serve (MI6) to St John Soanes Museum in London (Cabinet Office, 2022).[31] These sort of public bodies are required to advise government on what are sometimes very technical,

---

[29] Hancock, House of Lords Select Committee (2018: 354); Franklin (2013). Similarly, O'Neill describes the HFEA as having 'earned high respect' (2002: 127).

[30] Central government arm's length bodies (ALBs) consist of executive agencies (EAs), non-departmental public bodies (NDPBs) and non-ministerial departments (NMDs). EAs are business units within departments responsible for undertaking executive functions (Cabinet Office, 2016).

[31] See Cabinet Office (2023) for more detailed information on various shapes and sizes of public bodies in the UK.

complex and controversial issues.[32] Such is the case of the HFEA. With a special focus on regulating services and research involving any human oocytes or embryos, like all public bodies the HFEA is charged with protecting the public interest.

The HFEA's roots lie in the Warnock Committee of 1982 led by Baroness Mary Warnock, a Cambridge philosopher. The Committee's remit was to consider 'recent and potential developments in medicine and science related to human fertilisation and embryology; to consider what policies and safeguards should be applied, including consideration of the social, ethical and legal implications of these developments; and to make recommendations' (Warnock, 1984: 4). This came at a time when public anxiety was growing around the emergence of new reproductive technologies culminating in the birth of the first child, Louise Brown, in 1978 using IVF, a reproductive technology developed by Robert Edwards and Patrick Steptoe in Cambridgeshire:[33] 'As our reading of the evidence showed us, feelings among the public at large run very high in these matters' (Warnock, 1984: 1).

Warnock's report took two years to complete, and, in 1984, resulted in sixty-four recommendations on topics ranging from IVF, oocyte dona-tion and surrogacy to the regulation of embryonic research and human cloning. Despite the highly controversial nature of these recommenda-tions at the time, all were eventually incorporated in some form or other within UK law (Franklin, 2019a; Harding, 2024), including the passing of the Human Fertilisation and Embryology Act 1990 which, in turn, estab-lished the UK's HFEA in 1991 – the first regulator of Repro-tech in the world.

The HFEA is a statutorily independent NDPB, meaning that it operates at arm's length from government but ultimately is answerable to minister-ial scrutiny through its host governmental department, the Department of Health & Social care. One of the HFEA's primary functions is to license, monitor and regulate provision of all fertility treatments (such as IVF, gamete and embryo freezing) and any research involving the use

---

[32] See, e.g., www.gov.uk/government/organisation.

[33] Edwards and Steptoe performed the first successful IVF procedure at the Bourn Hall Clinic. Edwards was later awarded the Nobel Prize for medicine in 2010.

of human gametes and embryos, including pre-implantation genetic diag-
nosis and genetic editing in the UK as well as the prohibition of a range of
research areas such as the implantation of non-human embryos or
gametes or admixed embryos (part human, part non-human embryos)
into humans. As set out in the 1984 Warnock Report, the call for such a
public body rested on the understanding that 'public concern about the
techniques ... [involving gametes and embryos] needs to be reflected in
public policy', that 'the protection of the public, which ... [is] the primary
objective of regulation, demands the existence of an authority independent of Government, health authorities, or research institutions' and that,
crucially, such a body must not be 'unduly influenced by sectional interests' (Warnock, 1984: 75–79).

Franklin (2019a: 746) provides a fascinating account of the workings
of the original Warnock Committee set against the UK's 'pragmatic' legal
system, which she describes as not grounded in absolute ethical principles as in Germany or the United States but rather flexible enough to
permit a 'strict-but-permissive approach to promoting a science-positive
political (and entrepreneurial) context for controversial bioscientific
experimentation, which is that in exchange for allowing such research
to continue, it will be subject to the very highest level of regulatory
surveillance, directly overseen by Parliament and backed up by statutory
penalties for misconduct upheld through criminal law' (747).

As Franklin (2019a: 772) argues, Warnock showed how 'clear lines'
can be established on the basis of genuine debate 'leading to an agreed
set of limits and conditions, and that these limits and conditions can also
be changed over time using the same process' (772). Similarly, Harding
(2024) describes what came to be known as the 'Warnock Consensus' as
able to bring together opposing views through a structure of evidence-
based deliberation which gave the British Government the confidence to
accept even the most controversial of Warnock's proposals – that embryonic research ought to be permitted up to a strictly regulated threshold
of fourteen days.[34] As Harding explains, this was an extraordinary regulatory and political achievement that:

---

[34] Anne McClaren, one of the Committee's member and an embryologist, explained that
the primitive streak forms around the fourteenth day and is when cells combine to form

built public trust while allowing innovation to flourish. Democracy, civic participation and sensible governance led to the establishment of a deliberative committee that conducted research and listened to people. It was thoughtfully and expertly chaired by a non-scientist, Baroness Mary Warnock, who then spent years patiently explaining the compromise position they had reached. There was nothing glamourous about this process ... But it's the kind of steady, thoughtful democratic experiment which can bring scientific innovation and the body politic together. (2024: 76)

Aside from licensing and regulating research institutions and fertility clinics, another core function of the HFEA is to provide guidance to government for legislation, policy and the dissemination of reliable information on the usage of Repro-tech and related research. This is a vital public service that, while in consultation with the private sector it regulates, is nevertheless intended to be independent of its profit-orientated objectives.

Recent examples of the HFEA's ability to build public trust and convince government to adopt pioneering stances on the usage of new Repro-tech developments include the ethically sensitive decision to permit, under strictly regulated conditions, the technique called mitochondrial donation treatment (MDT) for hereditary mitochondrial disease. The majority of the 20,000 human genes are packed in the nucleus of most of the body's cells. Surrounding each nucleus are thousands of mitochondria with their own genes. These mitochondria provide the vital energy for the cells that develop into organs, almost like tiny batteries. When the mitochondria (inherited from the mother) is dysfunctional however, those particularly demanding tissues such as the brain, heart, muscles and liver tend to deteriorate as a child grows. Mitochondrial disease, which affects approximately 1 in 6,000 babies, can be a devasting and indeed fatal condition, that has many potential elements including substantial developmental and cognitive disabilities as well as major organ failure. Mitochondrial donation treatment is an

---

the beginnings of an individual: 'we can for the first time recognise and delineate the boundaries of a discrete human entity – an individual' (McClaren quoted by Harding (2024: 96)).

extraordinary procedure whereby two eggs are fertilised by the father's sperm – one from the prospective mother carrying the mitochondrial condition and on from an egg donor who does not. The nuclear material of the donor's fertilised egg is removed and destroyed. The nuclear material of the prospective mother is removed from their fertilised egg and transferred into the fertilised donor egg. This egg, with a full set of chromosomes from both prospective parents and healthy mitochondria from the donor, is then implanted into the prospective mother's womb and hopefully produces a perfectly healthy baby.[35] This will mean that there are three adults genetically involved in any resulting child (what the press donned as 'three parents'[36]) but only a tiny amount – about thirty-seven genes, or 0.2 per cent of the total – from the egg donor. Permitting the use of this extraordinary technique was a contentious recommendation, but such was the confidence in the HFEA that it had conducted a constructive consultative process with both experts and interested public parties and, most importantly, was competent in rigorously regulating such techniques in the UK, that the Government translated the HFEA recommendations into law in 2015[37] and the UK became the first country in the world to permit mitochondrial donation techniques. As the Chair of the HFEA at the time, Sally Cheshire, stated: 'This is an historic moment for the UK as we can now give women with serious mitochondrial disease the chance to have their own healthy genetic children for the first time' (HFEA, 2015).

Similarly, the HFEA's decision in 2016 to permit the Francis Crick Institute in London to use the genome editing technique CRISPR-Cas9 in human embryos under strictly regulated (*in vitro*) conditions was a historic regulatory decision. As the HFEA stated:

[35] The procedure is not without risks. Sometimes a tiny number of mitochondria are transported into the donor egg and multiply when the baby is in the womb (Dagan Wells interviewed in Sample 2023).

[36] 'Three-parent babies' claimed by Poulton (2016), who also commented on the first child ever born using MDT in 2016 in a procedure concluded in New York and not subject to US regulation.

[37] The Human Fertilisation and Embryology (Mitochondrial Donation) Regulations 2015 Available here: www.legislation.gov.uk/ukdsi/2015/9780111125816/contents .

These groundbreaking developments have happened *because of the regulation not in spite of it.* The Human Fertilisation and Embryology Acts 1990 and 2008 and the HFEA have provided a stable, yet flexible, framework in which UK bio-science and clinical expertise have been able to flourish. Scientists and clinicians have been able to go about their work free of the 'culture wars' that have hampered such activity in the USA or the regulatory free-for-all of much of the Far East. (HFEA, 2017: 2, emphasis added)[38]

The point to be made here is that good regulation cannot be said to stifle innovation, rather it is a vital part of it. Indeed, Franklin describes the HFEA as 'nothing short of a national treasure' (2013: 4). The technological ability to harvest human eggs, screen, alter and fertilise them outside a woman's body before implantation has not only changed the parameters of human reproduction but has expanded the research and future market possibilities of genetic engineering, cloning and even human–animal chimeras (Franklin, 2013). These, without any doubt, will have profound transformative effects on human life in the future and surely must be the subject of vigilant and well-resourced regulation in the interest of the public.

Given the importance of the HFEA, it was a shock to many to find that it had, in fact, been identified under the 2011 Public Bodies Act to be abolished, and remains under review.[39]

We said we would increase transparency and accountability, cut out waste and duplication, and we have ... the quango state will never again be allowed to spiral out of control. (Maude, 2011a: 2)

---

[38] As discussed earlier, science has more recently reached a new point of sophistication in that the first 'synthetic embryo' has been created. This development brings the creation and perhaps treatment of living creatures into ever sharper focus. The possibility of growing a human *in vitro* is not yet possible but is looking increasingly likely in the future. Natural embryos are currently fully covered by existing regulation; so much work needs to be done to catch up with this latest development (Weatherbee et al., 2023).

[39] The intention was to 'divide the HFEA's functions between a new research regulator, the Care Quality Commission and the Health and Social Care Information Centre' (Department of Health, 2010).

'Quango' is the derogatory name for public bodies (IfG, 2010)[40] which operate beyond the direct control of elected politicians and have been charged with creating a 'democratic deficit' in virtue of the power and public resources that they wield some distance from governmental departments (Levi-Faur, 2011; Dommett and Finders, 2015). However, while 'much maligned' as 'unaccountable, profligate and bureaucratic' (Dommett et al., 2014: 134), regulatory public bodies at the national level are considered by many as a 'necessary evil' (Levi-Faur, 2011: 15) as the nature of regulation becomes more complex and demanding (due to, e.g., hi-tech developments or increasingly complicated international legal compliance laws).

Prior to 2011, it was a commonly held cross-party view in Westminster that the UK, with over 900 regulatory public bodies, had become a 'quango state' and governance reform was urgently needed (Gash and Rutter, 2011).[41] Finally, the Conservative-led Tory–Liberal Democratic Coalition Government (2010–2015) set up a review into regulative public body reform with Francis Maude (Conservative Cabinet Office Minister) at the helm. Drawing on the dominant language of regulation scholarship, the aim of the review was set 'to reinvigorate the public's trust in democracy and also ensure that the Government operates in a more efficient and business-like way' (Cabinet Office, 2010: 1).

Maude promised radical change, and the eventual result was the Public Bodies Act 2011 which empowered ministers to abolish or reform public bodies in the UK. In order to guide ministers' decisions on the question of whether or not a given public body should be permitted to survive, Maude designed three tests: 'A body should only exist as a quango if it meets one of three tests ... 1. Does it perform a technical function? 2. Do its activities require political impartiality? 3. Does it need to act independently to establish facts?' (Maude 2011b: 1–2). These tests

---

[40] See Cabinet Office (2016) for full description of the various categories of UK public bodies (or ALBs).

[41] See also Dommett et al. (2014: 9), e.g., who explain how the abolition in 2003 of quinquennial reviews used to review ALBs 'was seen to have created a governance vacuum in which neither departments nor central government were consistently regulating public bodies'.

remain the same today in the context of the Cabinet Office's view that public bodies should only exist as a last resort (Gill and Dalton, 2023).

While these tests present seemingly sensible questions to ask of a public body, as a mechanism for public body reform, they gave rise to an unprecedented consolidation of power by Government to reshape the regulatory landscape of the UK in that it enabled ministers, with little to no oversight, to use the results of the tests to bypass parliamentary debate before abolishing, merging or modifying public bodies, many of which had come into being as a result of extensive parliamentary debate and deliberation. As Harries explained: 'If Parliament has thought this area so critical that it was worth weeks of its time to set up a regulatory body with very tight regulation in place, it hardly seems responsible to dismember that body with one quick snip and without serious consideration of the implications of so doing' (2010: Column 111).

By 2015 the Cabinet Office of the UK Government reported that through its Efficiency and Reform Group (ERG) which is charged with saving money and transforming the way public services are delivered through the Public Bodies Act 2011, it had already accomplished the significant saving of £3 billion by radically reducing the number of public bodies (Cabinet Office, 2015: 3). Indeed, 290 had been abolished, a further 165 public bodies were merged into fewer than 70 and the functions of over 50 further public bodies were moved out of the public sector, 'into innovative new models' (Cabinet Office, 2015: 4). Such models chimed with the ERG's double entendre slogan 'Government is open for business' (Cabinet Office, 2017) and several public body functions were moved to the private sector.

In line with these objectives, the UK Government introduced considerable legislation to complement the Public Bodies Act 2011,[42] all with the aim of reducing regulation further still but through complex and much more subtle means than Maude's public body tests. These measures were to fundamentally change the telos of the HFEA. Through the Enterprise Bill 2016, the Government emphasised its aim of 'extending the scope of our deregulation target to cover the actions of regulators, going further than ever before to tackle troublesome red tape'

---

[42] Available here: www.legislation.gov.uk/ukpga/2016/12/contents.

(Department for Business, Innovation and Skills, 2016: 1). Under this Bill, the Business Impact Target (BIT),[43] which requires ministers to assess the economic impact of regulators' activities on businesses, was set out. Then came the Growth Duty. First introduced in the Deregulation Act 2015, the Growth Duty places a specific requirement on regulators to 'have regard to the desirability of promoting economic growth in the exercise of that [regulatory] function' (ss 108 and 110, Deregulation Act 2015). As the Department for Business, Energy and Industry Strategy[44] reported, '[T]he growth duty clearly establishes Government's expectation that economic growth is an outcome that all regulators should be working towards' (2016: 1), and regulators must build an approach to delivering regulation that can 'assist in creating a more dynamic business environment, supporting innovation' (7).

Together these measures remain a key part of public body functions today (HFEA, 2023c), requiring regulators to continuously document the views of businesses on the effect that regulation has had on business capabilities and include these views in their performance reports. These, in turn, became a key feature in the review of each regulatory public body which must provide a robust challenge of the continuing need for individual NDPBs, both their functions and their form and where it is agreed that a body should remain as an NDPB, to review and ensure that the public body is complying with recognised principles of corporate governance (Cabinet Office, 2023). The tone of the reviews is somewhat threatening and the (ministerial level) reviewer is required to provide evidence and rationale for maintaining public body status for each regulator. In the words of Maude, the reviews 'not only make sure that quangos whose functions are no longer needed do not remain, but the reviews will also encourage bodies to explore new models of delivery and to drive through even more efficient ways of delivering public services' (2011a: 1). In effect, in order to survive, a regulatory public body must show not only that it does not hinder business interests but also how it

---

[43] First introduced in the Small Business, Enterprise and Employment (SBEE) Act 2015, which did not originally cover the HFEA.

[44] The Department for Business, Innovation and Skills became the Department for Business, Energy and Industrial Strategy in July 2016 when the new prime minister, THeresa May, took over as Leader of the Conservative Party in government.

actively facilitates the private sector. This sort of approach, now common in regulatory landscapes of democracies, raises very serious questions about the state's ability to design public policy with the public interest at its heart.

The majority of the HFEA's funding (approximately 80 per cent) is generated from the treatment and licence fees it charges the clinics and research establishments it regulates (government funding through the Department of Health & Social care. provides the rest).[45] It is required, on the one hand, to generate more funds from clinics and research institutes but at the same time to demonstrate how it is not a financial burden on businesses. In the Government's 'call for evidence' of the performance of the HFEA, one of the questions posed was: 'Could the Authority do more to support innovation and new approaches in the area of human fertilisation and embryology?' In answer, the Nuffield Council on Bioethics stated:

> No ... As a regulator, the HFEA's role should not be to promote innovation and new approaches, since this raises a potential conflict with its essential purpose ... Insofar as the HFEA should be subject to any duty relevant to research and innovation, it should be to promote public good consistent with public morality and the protection of the interests of patients and their off-spring, not to promote research and innovation *per se*. (Department of Health, 2016: 6)

Some critics welcomed the demise of the HFEA (Horsey, 2015). Savulescu, for example, surmised that we, as a society, should not attribute special moral status to the human embryo at all (a principle the HFEA remit is built on) and that accordingly the HFEA's attempts to work within 'conservative morality' (2011: 1) hinders both scientific and social progress.[46] However, the Government's rationale for the abolition of public bodies such as the HFEA was that they presented unnecessary red tape to thriving industry. In the particular case of the HFEA,

---

[45] For the financial year 2022/23, the Department of Health & Social care. contributed £1.198 million, and private treatment and licence fees generated £5.843 million (in addition to smaller amounts of other generated income) (HFEA, 2023c: 32).

[46] Also see Savulescu (2020).

passionate political debate about its future ensued across the Houses of Parliament and in public, and the HFEA survived, albeit its functions fundamentally reframed through the Regulator's code, the BIT and the Growth Duty.[47] Despite the HFEA's lucky escape and like all other remaining public bodies, it continues to be subject to a substantive review orientated towards market facilitation, and the possibility of its functions being transferred to the private sector remains open under the terms of the Public Bodies Act. Indeed, the HFEA must regularly demonstrate its alignment with market interests or risk abolition. Of particular note here is the argument that there is no requirement on the private sector to be transparent about 'the public money they receive, since this is regarded as commercially sensitive ... The contracts government has with private companies are private' (Dommett et al., 2014: 141).

I suggest that the 2011 Public Bodies Act was ironically far from effective in addressing any 'democratic deficit' that public bodies had come to be associated with, but it did serve to reduce the possibility of democratic debate on some of the most important national regulatory functions through its tight control of the HFEA's focus and the constant threat of abolition. As Judd argued:

> The Government keep emphasising that they want to improve the quality of democracy and to reduce the size of what they like to call the state. They say that they want to bring power and decision-making closer to the people. How such concentration of wide-ranging powers in the hands of Ministers will help to achieve those objectives is to me – and, I think, to most people – a total mystery. (2010: Column 141)

Indeed, considering the claims that these reforms served to reinvigorate 'the public's trust in democracy' and 'increase transparency and accountability' around regulatory practices, it was in fact more likely that they delivered much weaker democratic practices and the potential for far more opacity.

Aside from the continuing and explicit requirements to facilitate the private sector, another element of the HFEA's role in protecting the public interest is its public consultation. The HFEA prides itself on being

---

[47] See, e.g., HM Government, 2015: 57. 'Human Fertilisation and Embryology Authority'.

a public body with substantial expertise in public dialogue and consultation. There is no doubt that the HFEA puts a great deal of effort into reaching the wider public but it is bound by limited methods and there is, in fact, very little of the public in the actual process. One such example is the recent public consultation on the legal length of gamete and embryo storage, which has just been changed from a maximum of ten years to fifty-five years. As requested by the Department of Health and Social Care, the HFEA invited the public and any interested party to write to a specially designed website with their views.[48] The website set out both the legal history of gamete storage limits and the reasons why a new legal limit of fifty-five years might be better. One of the principal reasons for change was the topic with which I began this chapter – proactive egg freezing (HFEA, 2020). As the consultation document set out:

> Increasingly, people are choosing to freeze and store their gametes. This maybe because they are not ready to have a family but want to preserve their fertility so that they can start a family later in life ... Eggs frozen in a woman's 20s are at their peak of fertility meaning that, in general, the earlier a woman freezes her eggs, the better chance she has of achieving a healthy pregnancy. If frozen at this optimal age, a 10-year storage limit means that storage will expire in the woman's 30s, which may be too early for some people in the family-making decision cycle.

This was a very significant decision that, if taken up in great numbers in the future as Djerassi predicted, will have long lasting social and structural effects, not least in future familial structures. We are yet to understand the consequences of such a decision but a shift to progressively older-aged parenting is certainly possible. It will also require significant infrastructural investment in storage facilities to ensure that embryos and gametes are securely stored for very long periods of time, including beyond the lifespan of those individuals from whom they came. In response to the HFEA's guidance, the law changed in July 2022 to extend the general period of egg and embryo storage to fifty-five years (HFEA, 2020). Whatever the pros and cons of this decision, there was no publicly

[48] See the public consultation page on length of gamete storage at the Department of Health and Social Care (2021).

available data on the public consultation itself. We are told there were 1,222 responses analysed by the Department of Health & Social care., but we do not know the interests of these groups, institutions, industry representatives or individuals, nor why they sought out the HFEA website in the first place.

The question of public understanding brings us to the question of the HFEA's membership and how it reflects public concerns. Like the original Warnock Committee itself, the HFEA's membership does not include lay members in the sense of 'ordinariness'. Despite the fact that the Warnock Committee members were seemingly varied in their religious and professional backgrounds, all were institutional leaders or specialists with some relevant experience or strong interest in the field of Repro-tech, and all were personally selected by the Government (Warnock, 1984: ii–iii).[49] Warnock had hoped that the new HFEA would have lay members at its core:

> We would ... envisage a significant representation of scientific and medical interests among the membership. It would also need to have members experienced in the organisation and provision of services ... If the public is to have confidence that this is an independent body, which is not to be unduly influenced by sectional interests, its membership must be wide-

---

[49] As Warnock wrote in her cover letter to the report for Government 'Rightly you chose a membership which encompassed not only the many professions with a concern in these matters but the many religious traditions within society' (Warnock, 1984: iv). The Warnock Committee members were as follows: [Chair] Dame Mary Warnock, Philosopher and Mistress of Girton College, University of Cambridge; Mr Q. S. Anisuddin, Legal Executive, Vice-President, UK Immigrants Advisory Service; Mr T. S. G. Baker QC, Recorder of the Crown Court; Dame Josephine Barnes, Consultant Obstetrician and Gynaecologist; Mrs M. M. Carriline, former Vice-Chairman of British Agencies for Adoption and Fostering; Dr D. Davies, Director of the Dartington North Devon Trust; Professor A. O. Dyson, Samuel Ferguson Professor of Social and Pastoral Theology, University of Manchester; Mrs N. L. Edwards, Chairman of Gwynedd Health Authority; Dr W. Greengross, General Practitioner: Chairman of Sexual and Personal Relations of the Disabled.; Professor W. G. Irwin, Head of Department of General Practices, Queens University Belfast. Professor J. Marshall, Professor of Clinical Neurology, Institute of Neurology, London; Professor M. C. Macnaughton, Professor of Obstetrics and Gynaecology, University of Glasgow; Dr A. McLaren, Director, Medical Research Council Mammalian Development Unit; Mr D. J. McNeil, Solicitor, Edinburgh; Professor K. Rawnsley, Professor of Psychological Medicine, Welsh National School of Medicine; Mrs M. J. Walker, Psychiatric Social Worker.

ranging and in particular the lay interests should be well represented. We recommend that there should be substantial lay representation ... and that the chairman must be a lay person. (1984: 76)

It is true, of course, that the definition of 'lay' is open to interpretation, but this is an important feature that is intended to connect the workings of the HFEA more directly to the ordinary considerations of the 'everybody' as Arendt put it.[50] Thinking back to the discussions of the need to enable plurality to flourish in political deliberations in Chapter 1, this proposal for the inclusion of lay members by Warnock holds particular interest for me and I shall return to this topic further on. Suffice it to say here that while the membership of the HFEA consists of impressive professional individuals, its links to the public are, in reality, muted.

REPRO-TECH AND THE LIMITS OF LIABILITY

Warnock understood the intense public unease around embryonic research. She understood that such a fundamental issue went beyond party politics and urgently needed a different approach, one that built trust through the consensus of different views and interests with an outcome orientated firmly towards the public interest rather than a gladiatorial-style political brawl. Sarah Franklin's 2019 obituary of Warnock in *Nature* captures the essence of her work, which resulted in 'Britain's tightly regulated and legally binding, but popular and open-minded, system of biogovernance' (Franklin, 2019b). Writing later that year, Franklin is right to argue that the Warnock Consensus demonstrated that 'expert knowledge and reliable data are essential but never enough to enable enduring, humane governance to emerge' (2019c: 630).

Thinking of the shortcomings of AI governance (discussed in Chapter 2), Harding suggests that governance of AI ought to follow that of Repro-tech:

Much of the hyper-rationalist technology community would sneer at the concept of 'about right' in regulation and balk at the notion of including those with nonscientific backgrounds in a committee tasked with creating

---

[50] See Chapter 1.

those boundaries. But there is no other way for a democracy to approach breakthroughs that might alter society in fundamental ways by radically shifting the balance of machine intervention in our lives. (2024:118)

While the HFEA remains a gold standard public body in the context of the UK regulatory landscape and international comparators, I have tried to show here that, nevertheless, Repro-tech governance has been seriously limited by several factors. The first is its financial structure, such that it is dependent on the successes of the private sector in order to generate its annual budget. The second is that the HFEA is required to prove how it has actively facilitated economic growth in the field of human fertility and embryology. Third, only interested parties serve as a proxy for public consultation, and the membership of the HFEA is furnished by professionals with field-specific expertise but there is no 'ordinariness' to it. Fourth, the remit of the HFEA is limited to questions of liability (has illegal research been conducted?, was there wilful mis-advertising of IVF services?, etc.) and with very little scope for wider structural questions. Both Djerassi and Nozick were right in their predictions on genetic technologies – at its core are private interests, those of industry and those of parents. It is private interests such as these that have, if unchecked, the capability to accumulate to vast structural injustices grounded in genetic supermarkets and must become the focus of our political attention. While the current take-up of these technologies is still relatively small, there is no doubt in my mind that the future reproductive habits of generations to come will look very different in wealthier economies with some potentially dramatic consequences. As I discussed in Chapter 1, vast arrays of private interests culminate in habits and expressed beliefs and go on to form thoughtless structural actions that, despite being morally and legally acceptable in a given time and place, nevertheless become imbricated in the background conditions of macro-level structural injustices, and in turn shape the opportunities and life prospects of so many. Certainly, the potential for Repro-tech to become a significant element of the background conditions of structural injustices of the future is all too possible. Becoming politically responsible for the structural impact of Repro-tech requires a great deal more than a narrow liability-based regulation, especially one so strongly tethered to private interest growth.

CHAPTER 4

Putting the Public into the Public Body

Many [tech] researchers ... think that we are plunging toward a catastrophe, with more of them daring to say it in private than in public; but they think that they cannot unilaterally stop the forward plunge, that others will go on even if they personally quit their jobs. And so they all think they might as well keep going. This is a stupid state of affairs ... and the rest of humanity ought to step in at this point and help the industry solve its collective action problem

Yudkowsky (2023)[1]

To stay within the framework of Realpolitik is, I think, to accept a closing down of horizons, a way to seem 'cool' and skeptical at the expense of radical hope and aspiration.

Judith Butler (2020)[2]

TRANSFORMATIVE TECHNOLOGIES SUCH AS AI AND Repro-tech are set to profoundly change the way we live and organise social, economic and political life in the future. The context in which these technologies are designed, marketed and used, and the consequent impact they have on society and environment, ought to be of central political concern. As we have seen, while there are a plethora of tech-focused writers who warn us of a bleak future, very few attempt to

---

[1] Yudkowsky is a decision theorist, and is the founder of and lead researcher at the Machine Intelligence Research Institute in California. He has been working in this field since 2001 and is acclaimed as an AI expert. I am minded here of Elon Musk's recent pronouncement: 'If I could press pause on AI or really advanced AI digital superintelligence, I would. It does not seem like that is realistic so X [formally Twitter] AI is essentially going to build an AI ... in a good way, sort of hopefully' (Musk 2023).

[2] Butler's observation here echoes that of Arendt's call to think beyond the current moral and legal norms of the day to new futures discussed in Chapter 1.

engage with forging new practical political approaches. Suleyman (2023), co-founder of Deepmind, is amongst the exceptions.[3] He convincingly describes several elements of liability-based tech governance that he sees as essential going forward: these include, for example, corporate ethics boards that must include independent experts to develop self-regulating measures; and international alliances on regulating germline editing and autonomous weapons following the successes of other regulated technologies such as the Treaty on Non-proliferation of Nuclear weapons[4] and the Biological Weapons Convention on the prohibition of chemical weapons.[5] Another example is the very tight regulatory controls on pharmaceutical trials and sales around the world. Suleyman calls for state-level 'red-team' auditors (who proactively search for harmful elements and errors in AI and software) and greater state ownership and engagement with research and development of transformative technologies so as to ensure a key role for governments on overseeing technology's fast progress and deployment. Suleyman also advocates for governments to tax tech companies much more highly. He too is keen to see ordinary people central to decision-making and calls for essential 'public input'. Like Young, Suleyman thinks that such input is likely to come through social movements. Here, I want to explore how we might take a different route, building on some of the strengths to be found in the public body model for Repro-tech discussed in Chapter 3 but orientating the function of the regulatory landscape to operate with a broader remit than a commercial realpolitik limited to a

---

[3] Stuart Russell, an AI professor at the University of California; Berkeley is another.

[4] The Treaty on Non-proliferation of nuclear weapons (1970) 'is a landmark international treaty whose objective is to prevent the spread of nuclear weapons and weapons technology, to promote cooperation in the peaceful uses of nuclear energy and to further the goal of achieving nuclear disarmament and general and complete disarmament... A total of 191 States have joined the Treaty, including the five nuclear-weapon States.' United Nations Office for Disarmament available at the following link: https://disarmament.unoda.org/wmd/nuclear/npt/#:~:text=The%20NPT%20is%20a%20landmark,and%20general%20and%20complete%20disarmament.

[5] The Biological Weapons Convention (1972) 'prohibits the development, production, acquisition, transfer, stockpiling and use of biological and toxin weapons. It was the first multilateral disarmament treaty banning an entire category of weapons of mass destruction (WMD).' United Nations Office for Disarmament available at the following link: https://disarmament.unoda.org/biological-weapons/

liability-based approach to governance. Instead, I argue that we need to develop a different understanding of public interest that includes a focus on wider structural concerns and enables a reweighting of private and public interests in policy-making through lay-governance mechanisms. First however, I consider the status quo.

STATE FAILURE TO ADDRESS STRUCTURAL INJUSTICE

Young's theoretical work on the structural dynamics of everyday life provides what I see as an invaluable perspective on how we might begin to think about approaching the political challenges of addressing the structural dynamics of transformative technologies. While I have defended Young's view that structural injustice requires a different sort of political response to that which is motivated by liability, I find her account of structural injustice insufficient as a means to addressing the complex, multitudinous and cumulative actions of institutions, groups and individuals pursuing legitimate self-interests. Referring back to Chapter 1, Young was sceptical, particularly in her later work, of the state's capacity to address structural injustice, and instead favoured civil society–based social movements and public protest as an alternative approach. Recall Young's view on the state's tendencies:

> We cannot turn to the state or international institutions as arbiters in a struggle between the interests that produce structural processes with unjust outcomes and interests in changing those processes. The policies and programs that states and international organizations enact themselves tend more to reflect the outcome of those struggles than to balance between them or adjudicate them. (2011: 151)

Instead, Young thought that through 'vocal criticism, organized contestations, a measure of indignation, and concerted public pressure', individuals, groups and institutions could take up collective political responsibility through social movements to bring people together to focus on structural injustice (110). Such approaches are essential political expressions of frustration and hope for change, and typically there is a great deal of expertise and knowledge both codified and situated within social movements that help to form new political logics. Recent examples of successful social movements include Extinction Rebellion and Black Lives Matter,[6] which have impacted the ways in which governments and industry respond to public calls of injustice. These sort of civil society approaches remain a vitally important mechanism of social change, but they are not sufficient.

Social movements tend to rely on methods of protest, or semi-legal methods of occupation and obstruction, which can be effective and sometimes necessary to bring widespread attention to a particular issue but are, nevertheless, often challenging to fund and sustain over long periods of time with directed effect. These methods also tend to be driven by a small number of committed activists who must galvanise larger numbers around 'big messages' but tend to struggle to bring those numbers into complex operative discussion on how to persuade governments to help their objectives. Famously, when asked what action should be taken to combat climate degradation, Greta Thunberg, perhaps the most famous climate change activist in the world, replied 'It is nothing to do with me' (Thunberg, 2021). This is, of course, not a statement about lack of knowledge, imagination or care but rather a reminder that our political and institutional leaders are the ones with the greatest power to coordinate large-scale governance mechanisms and structural change. However, if we look to the machinery of the state, scholars such as Bickerton and Accetti (2017) observe 'a growing concentration of power in the hands of a set of unelected "regulatory bodies", drawing their legitimacy primarily from their technical competence and administrative expertise', creating 'a depoliticized form of "technocracy", where what is at stake is the preservation of the possibility of politics itself'. Some describe this sort of approach as a 'market police' which protects the interests of the most prominent market actors (Bonefeld, 2017: 50–51). Similarly, Brown observes that 'the state's task of security conditions for markets grows more complex as the economy does, making technocracy essential and further demoting the value or even possibility of democratic participation' (2019: 83).

---

[6] Their websites can be viewed at https://extinctionrebellion.uk/ and https://blacklivesmatter.com/respectively.

This sort of concern is often articulated in the form of regulatory capture. 'Regulatory capture is the result or process by which regulation ... is consistently or repeatedly directed away from the public interest and toward the interest of the regulated industry, by the intent and action of the industry itself' (Carpenter and Moss 2014a: 13). Capture scholarship began in the United States with the likes of Huntingdon (1953) and Bernstein (1955), and later developed by Stigler (1975), Peltzman (1976) and Becker (1983). It was weighted towards the idea that 'policymakers are for sale, and that regulatory policy is largely purchased by those most interested and able to buy it' (Carpenter and Moss, 2014a: 9). However, since then the field has moved its focus to more subtle forms of capture, as famously demonstrated by Sir Mervyn King, Governor of the Bank of England, during the 2008 global financial crisis:

> One of the major problems in regulation in the last 10–20 years has been that of regulatory capture. By that I do not mean that people were bought off. What I mean is that the sheer weight of resources, time and legal efforts that was put in by banks to try to persuade regulators that what they were doing was compliant with the rules made life extraordinarily difficult for the regulators. (2011: 2)

This more amorphous form of regulatory capture comes in many variations. 'Information capture' occurs when regulators are bamboozled by the technical detail of the industry they regulate and thus become dependent on industry expertise in such a way that compromises their independent regulatory capacity (Wagner, 2010; McCarty, 2014). 'Economists' capture' refers to the situation when scholars are seduced by industry to favour its own interests (over those of the public), thereby creating a bias that in turn is repeated by regulators who rely on scholars' pronouncements (Zingales, 2014). 'Cognitive capture' describes a sort of 'group think' that regulators adopt because prestige is associated with a certain trend of thinking in 'cutting-edge' industry (Johnson and Kwak, 2010), or when regulators are encouraged to rely on problem-solving toolboxes that are much the same as those of the regulated industry (Needham, 2010). 'Cultural capture' (sometimes called 'social capture') occurs when regulators closely work and socialise with the industry they are responsible for regulating and succumb to their influence, creating a culturing process to the industry's advantage (Kwak, 2014). When cultural capture is used specifically for the purpose of stretching the interpretation of rules or promoting deregulation in the interests of industry, it becomes 'corrosive capture' (Carpenter, 2014). In sum, all of these capture approaches to understanding the failings of regulators have a primary feature in common – that regulators are vulnerable to being 'turned', or as Posner describes, 'subverted' by the industries they regulate (2014: 49). Giving power to the technocratic machinery of the state, then, is to be avoided wherever possible, under this view. Certainly, such accounts align with Young's scepticism about the state's ability to address structural injustice and suggestions that these sorts of state regulatory mechanisms are likely to exacerbate structural injustice rather than offer effective tools for addressing it.

Thinking back to the case of the HFEA in Chapter 3, it seems to me, however, that a different aspect of regulator vulnerability is intentionally and publicly created by government, which I think of as more of a 'regulatory gift', rather than regulatory capture. By 'regulatory gifting' I mean to describe a form of gift from government to industry through a particular kind of deregulatory reform at the state level.[7] As anthropologists of 'the gift' explain, gifts are rarely neutral but instead act as a display of aspirations for closer alliances or a calculated gesture intended to secure particular interests (see e.g., Schieffelin, 1980; Mauss, 1990; Strathern, 1990; Laidlaw, 2003; Peebles, 2010). In selecting (or designing) and presenting a gift, the gift-giver accrues an agentic status that requires engagement, and should the gift be of positive value, a relationship of reciprocity develops, not necessarily in material exchange but one in which the gift-receiver's status of being identified as worthy of the gift is reflected in their confirmation of the gift-giver's 'wise decision' to select them as the recipient. That is to say, there is a reciprocal affirmation through which the interests of each are directed and effected. The affirmation between gift-giver and recipient may be observable to others and the public visibility becomes a significant characteristic of the relationship between gift-giver and recipient, shaping roles,

---

[7] I have written on regulatory capture and gifting in more detail elsewhere (Browne, 2020).

declarations and the nature of the gift itself. The regulatory gift – a gift from government to industry in the form of restraint, reorientation, or reduction of regulation and public bodies – is not simply deregulation, but rather, a set of actions designed to align with the private interests of industry in the hope of enhancing investment and economic growth and simultaneously cast in the language of public interest.

I suggest regulatory gifting differs fundamentally from the regulatory capture narrative that is so often used to justify the view discussed in Chapter 3 which is that 'public bodies should only exist as a last resort' (Gill and Dalton, 2023). Indeed, the most recent UK Declaration on Government Reform published in June 2021 committed the Government to 'commence a review programme for Arm's Length Bodies and increase the effectiveness of their departmental sponsorship, underpinned by clear performance metrics and rigorous new governance and sponsorship standards' (Cabinet Office, 2021). As we saw, the potential of Warnock's original idea for a public body was greatly diminished by what I am calling here regulatory gifting.

Regulatory gifting, then, differs from regulatory capture as a threat to the public body function in several ways. The first is the direction of intent. Regulatory capture occurs through 'action and intent by the regulated industry' whereby 'policy is shifted away from the public interest and toward [private] industry interest' (Carpenter, 2014: 63), whereas the regulatory gift does not rely on any particular regulator's surreptitious action but rather comes in the form of government's general policy on reducing regulation, and is overtly presented as being in the public interest. The second distinction regards the public nature of the gift. Regulatory gifting not only establishes a reciprocal alliance between government and industry but also acquires a political significance in that the gifting is intended to inspire a positive reputation for government as one that 'understands the bigger economic picture best' by cutting red tape for growing sectors such as Repro-tech and AI.[8] It is pitched as 'helping industry to create new jobs and ultimately to grow a stronger economy'. Industry's response is to endorse and congratulate

---

[8] The UK AI market, e.g., is expected to grow to £803.7 billion by 2035 according to the US International Trade Administration (Forbes 2023: np).

the government on its decisions and in so doing, further consolidate the government's image of economic competence, which, in turn, increases the perception that the public interest is being met. Third, unlike regulator capture, the regulatory gift is openly presented as working for the public interest by addressing inefficiency, opacity and other common regulatory public body dysfunctions. The regulatory gift is purposefully dynamic, in that it changes the nature of the regulator, and its objectives are aligned with that of private interest through the state. Lastly, government has the agency through regulatory gifting to effect change on a macro scale across the entire regulatory landscape and in regulatory dialogue with the international sphere, whereas regulatory capture is concerned more with the opening up of gaps in the behaviour and remit of individual regulators.[9]

Regulatory capture and regulatory gifting both serve to demonstrate how state machinery can become so heavily weighted towards private interest that, as Young argued, there is seemingly little hope of addressing structural concerns (or even some liability-based injustices). Yet I suggest it need not be so. Indeed, within a democracy, the normative purpose of a regulatory public body ought to be the protection of the public interest on the grounds that constraint on private (especially business) interests is a positive practice that benefits society (O'Neill, 2002; Christensen, 2011). This function speaks directly to a fundamental feature of the background conditions of structural injustice – the weighting of private and public interests. As I shall suggest, rather than

---

[9] In addition to regulatory capture theories, contemporary public choice scholars such as Shamoun and Yandle (2016) and Zupan (2017) have focused on individual governmental insiders. Their work has highlighted cases where, e.g., governmental actors sought to align their interests with those of industry in order to seek profit or further their own personal interests, and in doing so reformed specific regulations at the expense of the public interest. Shamoun and Yandle (2016) claim, e.g., that a head of state could be personally responsible for withdrawing or imposing particular regulatory reviews that benefit certain important political supporters of their own political agenda. Bearing the distinctions between capture and gifting in mind, however, we can also see how the regulatory gift differs from public choice-based arguments. While these concepts have some elements in common with regulatory gifting, in that the reform is initiated through the agency of the government, they differ quintessentially in terms of how the activities of government officials are pursued covertly and operate on an industry-specific scale.

abandon the state as too vulnerable to regulatory capture or regulatory gifting to address structural injustice, rather, the state's workings could be fundamentally reshaped. As Goodin argued, '[t]here are good reasons for pursuing certain sorts of goals through some sort of coordinated, collective apparatus like the state … Responsibilities get collectivized simply because that is the only realistic way (or anyway, much the most effective way) of discharging them' (1998: 50). Similarly, Runciman (2023b) argues that the state has what he calls 'super-agencies' that other institutions do not have, such as the ability to borrow vast sums of money and also the ability to sustain projects over the long term to enable its objectives.

As we saw in the previous discussion of the HFEA, the public body has many operational features that are fundamental to the state's functioning. In what follows, I begin to think about how we orientate huge public body machinery away from both regulatory capture and regulatory gifting while retaining the state's ability to effect macro-level structural change through the enactment of political (structural) responsibility. I begin with the *raison d'etre* of the public body – to serve the public interest – before moving on to questions of who is best placed to speculate on the structural dynamics of transformative technologies and think through what I might mean to take up structural responsibility.

THE PUBLIC INTEREST AND POLITICAL RESPONSIBILITY

The very first call on those serving in public bodies is that they 'should act solely in terms of the *public interest*' (HM Government, 2023d).

At a minimum, Barry's observation that the public interest must correspond to 'those interests which people have in common *qua* members of the public' (1965: 190) surely must hold. However, despite being 'an established political and normative concept with great policy resonance' (Minteer, 2005: 38),[10] 'the public interest' is notoriously opaque. Many would find their view of the public interest reflected in the words of mid twentieth century political scientists Dahl and Lindblom who described the public interest as 'nothing more than

---

[10] See also, e.g., Sunstein and Vermeule (2020: 6).

whatever happens to be the speaker's own view as to a desirable public policy' (Galston, 2007: 11). Yet the concept lives on as one of the core objectives of democratic government.

The idea of the public interest is often used interchangeably with the 'public good' or the 'common good'.[11] Mansbridge and Boot (2022) provide a fascinating history of the intertwined usage of these concepts, and demonstrate their importance to political thought ranging from ancient Greece to the modern day. Loosely speaking, three theoretical interpretations are presented as being the most familiar today: aggregate, unitary and procedural. The aggregate interpretation sees the public as a number of individuals whose interests can be imagined as an aggregated common good which ought to be maximally sought (see, e.g., Bentham, 2007 [1780]) and for some theorists, this attempt ought to be constrained by the duty not to disadvantage those worst off (e.g., Rawls, 1971). The unitary interpretation presents a rather different view. Here the common good is understood to be a particular prescriptive goal that ought to be pursued even at the cost of some individuals: a 'sacrifice for the greater good' (Sandel, 2020: 112). Either way, aggregate or unitary, it is not clear who legitimately determines what the public interest ought to be and at what cost to some. Finally, the procedural interpretation sees the public interest or the common good as the outcome of a democratic deliberation where the views of the public are represented and each participant must try to see the world from others' perspectives (Habermas, 1989 [1962]; Benhabib, 1992). The obvious attraction of this approach is that it is not limited to any particular conception of the common good and moves beyond a simple interest-based conception of democracy whereby individuals and interest groups vote for outcomes that they perceive as best serving their own narrow interests. By contrast, the ideal of deliberative democracy promises citizens the opportunity to discuss and debate collective problems and objectives. Here we might think back to Arendt's argument that 'politics is based on the fact of human plurality' (2005: 93). Certainly, Arendt was

---

[11] I use the terms 'public interest', 'public good' and 'common good' to mean the same thing in this book.

against the idea of a doctrinal or fixed conception of the public interest or common good as the central focus of politics but she did appreciate that each of us has a particular perspective on the world, moulded by our own ideas and experiences, and thus none of us can objectively comprehend the world as it really is in all its complexity. Only through engagement and exchange of ideas and experiences with others can each of us begin to understand the diversity of our collective existence and its environment. This has consequences for a more responsive conception of the public interest that permits speculation as to the nature and shape of the background conditions of structural injustice. As Sluga argues, the idea of a common good or public interest 'cannot be found by individual thinkers withdrawing from the world. The search for it requires, instead, engagement in the world and with others' (2014: 159). Nevertheless, an obvious political question remains when we move from the abstract to the empirical: What political apparatus is needed to facilitate pluralistic speculation on the structural dynamics of transformative technologies? Certainly, this is not, as Walzer put it, 'an activity for the demos ... 100 million of them, or even 1 million or 100,000, cannot plausibly reason together' (1999: 68). Nevertheless, as Hill Collins argues:

> Because subordinated groups are routinely excluded from formal institutions of governance and knowledge-construction, the resulting social inequalities that they experience limit their ability to exercise power within and across multiple domains of power. This exclusion in turn limits effective problem-solving because the perspectives of the people who are most affected by social problems are silenced. (2017: 23)

Even where political institutions make great efforts to acknowledge particular experiences of intersectional injustices such as racism and sexism, each particular context will have a 'distinctive matrix of domination' (2017: 122–123), and thus the particularities of each cannot easily be reduced to the other. In a similar vein, Young called into question the 'assumption of unity' as either the precondition or the goal of democratic deliberation which serves to exclude those who do not fit (1997: 66). More to the point, Young is right to express scepticism that in a pluralistic society there are 'sufficient shared understandings to appeal to in many situations of conflict and solving collective problems' (1997: 66), and that a core problem with placing unity as the central feature of deliberation is that it 'obviates the need for self-transcendence': participants are merely orientated towards tolerance and compromise rather than a willingness to be open to a different way of conceiving of and solving collective problems together. If the focus is fixed on what people share then 'each finds in the other only a mirror for him or herself' (Young, 1997: 66).[12] This point takes us back to the predominance of the technocracy, whereby public bodies tend to be populated only with experts, industry-specific or other relevant professionals. Young's proposed solution, much like Arendt's, was to treat difference as an asset to understanding rather than an obstacle to flatten out in the quest for the public interest. This is an idea that has been shared with many scholars (such as, e.g., Butler, 2018)[13] and chimes with Mansbridge's argument that the contested nature of the idea of a common good or public interest is part of the unsettled and contested nature of politics itself:

> [it] undermines a philosophical agenda whose goal is settled knowledge, either of the self or of the polity ... Yet the moral import of choosing the common good has not changed ... one does not need certainty about the meaning of the common good to act for the common good rather than self-interest when trying to live ethically. Living with such uncertainty is required today if widespread commitment to the common good is to help solve pressing collective problems and generate the mutual trust necessary for efficient and more importantly for moral interaction. (2013: 10)

This sort of unsettled status of the public interest and indeed politics itself fits well with the processes of structural speculation that I discussed in Chapter 1, as well as the transient relationship between structural, moral and legal forms of political responsibility. Recall, that speculation requires us to look beyond what we think of as traceable evidence that individuals, groups or institutions are liable for structural phenomena. Nevertheless, such an approach remains open to the possibility that new

---

[12] See also Achen and Bartels (2017) who argue that bias is enhanced by democracy whereby voters merely seek out politicians whose views are most closely aligned with theirs.

[13] Alongside the question of how to translate the ideal of deliberative democracy into practice (Walzer, 1999; Dryzek, 2001; Achen and Bartels, 2017; Fishkin, 2020).

epistemologies may come in the future that enable the transition of some structural injustices to moral or legal forms of fault.

I also argue that because the political choices that we all face are so intertwined with the highly complex, diverse accumulated outcomes of others' actions and an inevitable conflict of values and objectives, a primary focus of structural responsibility for the public interest ought to begin with what I call a 'negative conception of the public interest'. What I mean by this is to begin by focusing explicitly on what is *not* in the public interest rather than with the requirement to first formulate universal conceptions of 'the good life' to guide political action. I agree with Shapiro (2016: 7)[14] on the point that the public are much more coherent about what they do not want than what they do want.[15] That said, even the negative public interest is set against a background of political uncertainty; as Dunn has argued, 'no sound understanding of politics can be founded on an essentially pre-political or extra-political conception of the human good ... [p]olitics depends on hypothetical imperatives all the way down' (1990: 195).[16] Similarly, Sluga argues: 'politics is not the implementation of a fixed common good; it is, rather, an ongoing search in which various conceptions of the good will be proposed and dismissed. That we engage in this search does not mean that there is a determinate good to be found' (2014: 4).

In concert with this view, I want to broaden the way in which public interest is used as the grounds for governance mechanisms such as the public body. Although the normative purpose of a public body is to protect the public interest, I demonstrated in Chapters 2 and 3 when focusing on the governance of increasingly transformative technologies,

---

[14] Also see Barry (1965).

[15] Here Shapiro rejects, as the grounds for political action, Kant's universalising concept of the categorical imperative: 'Act only according to that maxim by which you can at the same time will that it should become a universal law' (Kant, 1959 [1797]: 39), in favour of Kant's alternative hypothetical imperative (actions based on desired outcomes).

[16] This too we see echoed in the work of Weber (2002 [1919]), who argues that in politics, an ethic of conviction cannot be permitted to override an ethic of responsibility where it is imperative that consequences of actions, intended and unintended, are addressed head-on against a background of inevitable and sometimes irresolvable conflict of values and objectives.

AI and Repro-tech, that the 'public interest' has been reduced to a notion of hazard and risk grounded in liability that excludes crucial structural considerations. Here the public body's role in the mitigation of market failure looms large and the public interest all too often manifests as merely checks and balances on potential wrongdoing, such as unsafe practices, corrupt practitioners, false advertising, substandard products and discriminatory behaviours. The management of these is of course important as well as necessary to maintain market confidence but, I argue, achieves too little else in the public interest. I suggest that there is a failure to address the more complex structural harms emanating from the accumulated, unintended, uncoordinated outcomes of the structural actions of individuals, groups and institutions. Given that those who are economically vulnerable or socially marginalised tend to suffer structural injustice rather than elites or corporations, surely these structural dynamics are of the utmost public interest too and should be reflected as such in the functions of the public body. As Runciman (2018, 2023a) argues, many people feel ignored by legal and policy-making processes largely dependent on technocratic thinking, that seem no longer to be designed to benefit an increasingly unequal and diverse population.

Indeed, Runciman observes that current forms of representative democracy, largely dependent on technocratic thinking, fail to address the more nuanced concerns of a diverse public: 'For democracy to flourish it needs to retain its ability to combine net benefits with personal recognition' (2018: 214). Democracies then need to inspire their citizens to express themselves on vital political questions of the day. However, there are serious limits to public engagement with state policy-making despite best intentions, let alone speculating on the structural actions that may cause macro-level injustices.

Why then might we consider the state's regulatory public body machinery to be helpful for addressing the background conditions of structural injustice? How might a state public body model avoid the formulaic 'unity' that Young was so critical of whilst also facilitating pluralistic speculation about structural injustice to a productive end? This is a large question but part of the answer lies in the state's coordinating power at the macro level.

THE PERENNIAL PROBLEM OF PUBLIC
POLITICAL ENGAGEMENT

As Young argued, our responsibility to address negative structural dynamics 'is essentially shared with others because the harms are produced by many of us acting together within the accepted institutions and practices, and because it is not possible for any of us to identify just what in our own actions results in which aspects of the injustice that particular individuals suffer' (2011: 110). I have clarified these actions as 'structural actions' to distinguish them from all actions or lability-based actions and have argued that in engaging with these sorts of structural harms the political habit of pluralistic speculation beyond liability is required in our governance approaches. However, enacting this sort of responsibility on a large scale to coordinated effect is a great political challenge not least because of the perennial problem of low levels of public political engagement.

As Achen and Bartels point out, engaging the public in detailed political decisions is often unrealistic:

> Human beings are busy with their lives. Most have school or a job consuming many hours of the day. They also have meals to prepare, homes to clean, and bills to pay. They may have children to raise or elderly parents to care for. They may also be coping with unemployment, business reverses, illness, addictions, divorce, or other personal and family troubles. For most, leisure time is at a premium. Sorting out which presidential candidate has the right foreign policy ... is not a high priority for them. Without shirking more immediate and more important obligations, people cannot engage in much well-informed, thoughtful political deliberation, nor should they. (2017: 9)

I disagree with the notion that 'nor should' the public be engaged in the political questions of the day, but I take Achen and Bartels' main point that even for those who attempt to engage with the details of political debate before making a considered decision, most are only able to scratch the surface with little grasp of the wider context or future implications. This is what Downs called 'rational ignorance' (1957: 139). What tends to grow into the spaces created by rational ignorance is simplified binary politics – in or out, with us or against us, group versus group. These sorts of binary politics invariably lead us to new and renewed forms of division and segregation. We see, for example, a growing suspicion of diversity and inclusion agendas perceived as diluting national interests for the sake of woke objectives of plurality. As Foa and Mounk observe, there is an 'active embrace of illiberal movements hostile to pluralistic institutions' (2019: 1013). Populist politics is on the rise and often includes narratives of '"pure people" versus the "corrupt elite"' (Grzymala-Busse et al., 2020: 1) or 'harness disaffection and amplify otherwise legitimate concerns in ways that manipulate public will' (Niemeyer and Jennstål, 2018: 329). Fishkin crystallises the problem of political response: 'listen to the people and get the angry voices of populism or rely on widely distrusted elites and get policies that seem out of touch with the public's concerns. Populism or technocracy?' (2018: 3). Neither is a solution to the problems of designing public policy with a more structural remit in the democratic context.

One approach to solving this problem has been to diversify political representatives in deliberative democratic fora to better reflect a pluralistic population. Despite the populist tendency to 'represent the politics around gender or racial equality as yet another elite preoccupation' (Phillips, 2020: 185), the diversification of representatives remains a common objective of democracies.[17] This has led to some important progress: the UK Parliament reports, for example, that after the last general election in the UK there were sixty-six (10 per cent) members of the House of Commons who identified as being of a minority ethnic background (House of Commons Library, 2022)[18] and 225 (31 per cent) as women (House of Commons Library, 2023a). These are record numbers of change. However, it is also notable that 61 per cent were privately educated and 45 per cent of the Cabinet in October 2022 held a degree from Oxford or Cambridge University (Sutton Trust, 2022), suggesting less diversity than we might have assumed from observing the rise of other social characteristics. As Ahmed's (2012, 2017) work demonstrates, stated diversity objectives can be held up as 'evidence' that

[17] See, e.g., Reingold (2019).

[18] Against a backdrop of 14.4 per cent of the population identifying as minority ethnic.

institutions have done their bit without really addressing deeper institutional segregation patterns of exclusion, racism, sexism and homophobia, for example. This practice Ahmed calls 'non-performative' (2012: 117), which depicts policies that 'stand in for the effects' and become the ends in themselves rather than the means to real improvement. That is to say, to have a public-facing diversity policy becomes the objective itself rather than the substantive achievement of diversity.

As important as it is, diversifying our institutional representatives can only take us so far. Implicit in the classic work on political representation by Hannah Pitkin (1967) was the question, 'why should someone who resembles me be good at representing me?' The underlying thought here is that what really matters is that our political representatives substantively represent the people's interests, not the resemblances they share. However, what tends to happen if we bypass concerns about diversity – what Pitkin called 'descriptive representation' – whereby representatives share a range of characteristics with the pluralistic citizenry they are tasked with representing, is that those homogeneous groups which tend to dominate political assemblies as our representatives (traditionally white, middle-aged men with similar cultural and professional profiles) can simply claim they are substantively representing the people irrespective of their homogeneity, and thus the status quo of political representation is maintained. Certainly, there are 'male politicians who have fought tirelessly for women's rights; female politicians who disdain any suggestion that they might speak for women; and white people who have given their lives to struggles for racial equality' (Phillips, 2020: 187). But still, intuitively something is wrong with an outcome that maintains overwhelming homogeneity in political assemblies which sit against the backdrop of a highly diverse population. Thinking specifically about the tech context, Costanza-Chock (2023: 377) similarly describes the 'ongoing reproduction of white cis male "tech bro" culture that alienates women, trans* folks, B/I/PoC, Disabled people, and other marginalized communities' in technological development and governance.

An obvious alternative to non-pluralistic representation is based in essentialism. Representatives who are reduced to their gender, the colour of their skin, their religion or economic background as a proxy for understanding or predicting the views of groups with whom the representative shares a characteristic are also limited (as Pitkin's work highlighted). Many scholars have provided important insights on the conundrum of representation versus essentialism. Rather than focusing exclusively on representation, which, when looked at in detail, becomes a paradoxical concept (how could representatives ever accomplish the task of representing a diverse citizenry with a multitude of different views?),[19] scholars such as Phillips ask us to include a focus on presence: 'there can be no substitute for the presence of those with the more direct experience in decision-making assemblies' (2020: 177). Similarly attempting to avoid essentialism, Young called for a plurality of perspectives based on lived experiences: 'a perspective is a general orientation on the political issues without determining what one sees, and without dictating particular conclusions' (2000: 148). Here I am reminded of my earlier work on segregation (Browne, 2006, 2014). I sought to understand the arguments both philosophical and practical around 'solution policies' in the context of acute segregation patterns, such as quotas for women in the workplace as a way of diversifying institutions and countering perennial trends of gender inequality. I was not convinced that blanket quotas for women were the best way to approach this sort of injustice. Much of the debate at the time focused on the European Commission's proposal (draft directive COM 614), to place an 'obligation of means' on all listed companies to ensure that at least 40 per cent of non-executive directors or 30 per cent of all directors of each corporate board were female by 2020[20] – a policy designed to be terminated in 2028. The objective of the quota was to raise 'the representation of women' but the outcome of such a policy tended to create a loop effect where a relatively small number of women (or at best women with very similar profiles) were circulating through corporate boards as non-executives.[21] In many cases, this was an illustration of Ahmed's non-

---

[19] See Runciman (2007).

[20] There was no intersectional element to the policy but rather a simple binary of women versus men.

[21] At the time COM 614 was being introduced, I was invited to Brussels by the European Institute of Gender Equality to address an expert audience including members of the EU Parliament and the European Commission. I certainly sympathised with the lack of

performative practice. To think through the lens of a negative public interest, it is clearly against the public interest to have a segregated labour market where women are discouraged from leading institutions and where instead individuals are locked into rigid gender-based work roles according to social characteristics rather than ability. So too is it against the public interest to ignore patterns of associated lower pay which compounds discrimination and increased precarity for contemporary and future generations of individuals and families. In this sense, the increase in women on boards was a positive outcome. However, to use the language of Young, it is not clear that a plurality of perspectives capable of changing the structural relations within employing institutions was achieved even though the numbers of women were improved through the quota policy. In a similar vein, Fraser (2015: np) argues: '[f]or me, feminism is not simply a matter of getting a smattering of individual women into positions of power and privilege within existing social hierarchies. It is rather about overcoming those hierarchies. This requires challenging the structural sources'.

How then do we bring a pluralistic presence capable of speculating on diverse experiences of the background conditions of structural injustice to questions of governance? The common narrative that technocrats are out of touch with the people and populists distort the 'public interest' for their own ends has encouraged policy-making institutions to combine their efforts with more direct citizen engagement through public polling, focus groups or public consultation.

The current UK Government has been very fond of nationwide polls as a means of testing public opinion. Indeed, it had, for example, been

progress on women's unequal pay, which is currently at 8 per cent in the UK (ONS, 2023) and low levels of women in leadership roles. I agreed that in the absence of any other sort of mechanisms, blanket quotas were better than nothing. However, rather than focusing solely on elite layers of power in corporate institutions (the sole focus on COM 614), I suggested instead a wider sort of mechanism for addressing more structural institutional segregation patterns – the Critical Mass Marker Approach. This approach focuses on all levels of institutions or sectors and requires the setting of quotas according to the disproportionate movement of women from one level to the next, thereby ensuring movement up and across employment structures where critical mass had failed to manifest (Browne, 2014).

something of an obsession in Downing Street, as demonstrated by this quote by an unnamed government official:

> The internal polling [Downing Street] is pretty extensive every day ...
> We get an overnight breakdown of surveys of 2,000 adults. We get stats on
> how worried people are, people's perceptions of risk, whether they feel
> they are being served by government information, whether we have got
> the balance right between the economy and healthcare, polling on
> people's finances, thoughts on the NHS. (Cooper, 2020)

The way this sort of approach works is that a respected polling company, such as YouGov, will have about a million UK citizens on its books at any one time. These will have been reached through targeted advertising, and if accepted are paid a fee every time they fill in a survey on request. For each survey, a smaller number (such as 2,000) are chosen according to some basic social characteristics such as age, gender, socio-economic status and education, and, depending on the question, more will be added, such as place of birth, voting behaviour, education level, etc. Once the data is collected it is weighted against the national profile of adults (using the same sort of criteria) using, for example, large-scale datasets such as Census data, Labour Force Survey, general election data and ONS data. There is no doubt that public polling is a highly valuable exercise and certainly helps to access much more of the public's plurality when designing policy in its name.

However, as discussed earlier, these sorts of data give us an account of what we *already* think and there is little room for engagement with alternative views, debate or 'self-transcendence' (Young, 1997: 66), and for those filling in the surveys there is no encouragement to step back and look at the 'choice architecture'[22] of the options from which to choose nor indeed of the life decisions they have made or have yet to make. Indeed, YouGov's advertising slogan is 'Share your opinion, Earn money, Shape the news'. While this is an accurate description, it reveals an absence of speculation about one's structural relations to others, about evidence from alternative views as well as the stories of those who have experienced unjust structural outcomes.

---

[22] See the description of Haslanger's concept in Chapter 4.

A more complex polling approach comes in the form of deliberative polling or focus groups (Fishkin, 2018) whereby a small group of individuals are selected according to general characteristic criteria and the process begins with small group discussions, question and answer sessions with experts and moderators, and a confidential questionnaire which is passed on to the body concerned for consideration. While providing greater opportunity for discussion, these groups tend not to be furnished with the opportunity to debate with national- and international-level experts or have any direct impact on state-level decision-making. Moreover, those invited to take part are not recorded for the general public to follow, but tend rather to be small and private gatherings focused on a narrow set of policy ideas.

Although 'public consultation' is a common claim of government decision-making, it tends not to be very public. Opinion is solicited usually in the form of written material submitted by a range of different agents over a twelve-week period. Materials are then sometimes (but not always) published on a related government website for anyone interested to see. However, this opportunity only comes to those who are specifically looking for a particular consultation process on government websites (and even then they are quite hard to find, as I noted in Chapter 3). Such consultation processes tend to be organised around an information document about which the public can then write to the relevant committee through an online portal. Not surprisingly, these processes tend to be dominated by private interest groups and industry, and have very little engagement with the general public – that is to say, there is very little pluralistic presence in the data that is meant to inform governance decision-making. Here we can think back to the limits of public consultation on AI and Repro-tech regulation discussed in Chapters 2 and 3.

Attempting a structural focus in public policy is a difficult challenge. However, it is not sufficient to hope that the public will spontaneously develop a habit of structural speculation or invest enough time and commitment to educate themselves on all of the important political questions of the day – what Achen and Bartels (2017: 1) call the '"folk theory of democracy" whereby citizens are assumed to make coherent and intelligible policy decisions, on which governments then act'. Nor is it sufficient, as I have discussed, to diversify our political representatives (as vital as this element of democracy is), not least because all politicians also inevitably orientate their actions towards short-termism and internal party strategies which can get in the way of long-term public interest questions.[23] Nor is it enough to conduct political responsibility by public poll. As I discussed in Chapter 2, humanity has never been more connected than it is today through X (formerly known as Twitter), Facebook, WeChat and other social media fora, but the idea of an 'Xcracy' or a 'Facebocracy' is anything but exposure to and deliberation about different ideas, rather, it is an intensifying echo chamber populated by increasingly convincing deepfake materials. Neither does the answer lie with the very limited procedures of focus groups and public consultation which tend to inhabit a grey area around governments with little account of why certain members are chosen and the nature of the exercise and the parameters of what they are asked to consider.

What I suggest we need more of is a pluralistic lay-presence built directly into the mechanisms of democratic governance in such a way that enables deliberation and exposure to competing forms of evidence under specialised conditions and an obligation to speculate on the structural dynamics of everyday legitimate pursuits of private interest. As Dunn argued, 'What modern politics most pressingly requires is a *democratization of prudence*, a spreading out of the burden of judging and choosing soberly about political questions across the entire adult population of particular societies' (1990: 214; emphasis added). The drive to radically improve policies that serve the public interest is critical if democracy is to survive as a prominent form of governance (Runciman, 2018; Brown, 2019; Sandel, 2020), and I suggest central to this concern is a politics capable of thinking about the structural injustices of transformative technologies rather than only the market-enhancing liability approaches to governance.[24] Dunn's call for a democratisation of prudence must surely be right in some sense, but Achen and Bartel's 'folk democracy' is a persuasive challenge nevertheless.

---

[23] See, e.g., Greta Thunberg's speech on this point in the context of environmental degradation (Austrian World Summit, 2021).

[24] Or as the title of Taylor's (2019) book puts it 'Democracy may not exist, but we'll miss it when it's gone'.

Here, I find the work by Jugov and Ypi (2019) illuminating. Their focus is on the ways in which those on the sharp end of a particular structural injustice might become to be aware that their presence in the complex array of structural relations and power dynamics (or to use Hill Collins's description, 'matrix') play a part in perpetuating the background conditions of a structural injustice that oppresses them. As they argue, 'different political responsibilities might correspond to different degrees of epistemic awareness depending on agents' perception of and ability to reflect on the injustice they suffer within a structure' (2019: 12).[25] Even though the convoluted and complex elements of structural injustice are too obtuse to map onto individuals, groups or institutions in a meaningful way that could be traced to liability, the effects of structural injustice have a systematic or stable quality that serves to constrain certain individuals, groups or institutions. Young's view was that because structural injustices operate beyond the usual legal and moral frameworks for thinking about injustice, they are all too often disregarded as misfortune or bad luck. As Dillon and Craig (2021: prologue) describe, the acts of reasoning and judgement are always populated by everyday stories; '[s]tories, in their many shapes and forms, are part of human nature and humans are immersed in them: they are experienced as deeply individual and as integral to relationships between people; they provide explanations, meaning and entertainment: people die for them'.[26] Stories, in amongst the facts and the figures that give shape to any account of social life, provide a broader picture of the experiences of structural injustice and the complexities of its background conditions.

Despite the challenges that untraceability brings to taking up political (structural) responsibility, everyday stories and experiences provide a

[25] See also Young (2011: 113) 'those who can properly be argued to be victims of structural injustice also can be called to a responsibility they share with others to engage in actions directed at transforming those structures. Indeed, on some issues those who might be argued to be in less advantaged positions within structures perhaps should take the lead in organizing and proposing remedies for injustice, because their interests, it might be argued, are most acutely at stake.' Also Young (2011: 148).

[26] I am also reminded here of Rorty's (1995) argument that peoples' stories generate in others a sense of sentimentality and sympathy that enables a more collective approach to politics. Rorty's particular focus was that of human rights but the point holds for a wider remit of politics.

sense of the power dynamics that inhabit the background conditions of structural injustice and this I suggest is the most productive site of potential structural change. If we think back to the stories and experiences that illustrated the implications of AI decision-making in Chapter 2 and the individual motivations to gravitate towards genetic technologies in Chapter 3, we can start to see that governance of these technologies ought not to be solely grounded in who is at risk of malpractice or mis-advertising but also in the macro level implications that these technological shifts have for ordinary people across populations. As Young argued, rather than limiting ourselves to orienting political machinery to tracing culprits when that proves impossible, we should instead look to structural change. Unlike Young, however, who thought that, 'a state has too many interests in these structures to be politically responsible' for structural injustices (Beardsworth, 2015: n6), I see potential in a radical transformation of the public body model as a productive way forward at the state level.

I suggest that a lay-centric approach be built directly into the mechanisms of regulatory state governance with specific structural responsibility to speculate on the power dynamics of private and public interests in society, as well as on how everyday experiences feature and contribute within those power dynamics as part of the background conditions of structural injustice. Here I draw on some inspiration from an experimental form of political deliberation, the mini-public, which creates the opportunity for the public to speculate and give policymakers a much richer sense of what the plurality of experiences connected to structural injustices might be. There has of course been a great deal of scholarship on mini-publics[27] in different formats but none has been focused on the regulatory public body landscape in the way I shall come on to sketch.

THE MINI-PUBLIC AND ITS LIMITS

Although they loom large in the democratic literature, mini-publics are, in fact, rarely found as mechanisms of governance.[28] Nevertheless,

---

[27] See, e.g., Bächtiger et al. (2018), Lafont (2019) and Fishkin (2018).

[28] Robert Dahl is often identified as the first to discuss the idea of a mini-public for contemporary policy-making As Escobar and Elstrub explain: 'Inspired by democratic

discussion of their merits and limitations (mostly in contexts other than government) are enlightening. In an attempt to address concerns of the failing legitimacy of democratic institutions while sidestepping the over-demanding practical dimensions of ideal deliberative democracy where each of us represents ourselves, mini-publics all share the feature of 'deliberation concerning general political issues among small groups of randomly selected citizens' (Lafont, 2015: 40). The design of mini-publics comes in many different shapes and sizes (Fishkin et al., 2017; Setälä et al., 2021), but usually, members are chosen through stratified random sampling from the electoral roll so as to create a proxy for the given population including demographic characteristics such as age, gender, ethnicity, income, geographical location, religion, etc. The lottery element of selection 'underpins the legitimacy of the selection process' (Escobar and Elstrub, 2017: 1), which is designed to consider a set of issues that affect the whole population. Those laypeople selected from the general population who agree to become members of the mini-public are remunerated and professionally convened in a single location. Experts provide the participants with evidence and also advocate a range of positions which are then cross-examined by the participants. Discussions are facilitated by professionals who document the ideas and final votes into a report, usually focused on a single issue. The mini-public is dissolved as soon as the report is completed.

In the UK in June 2018, the House of Commons commissioned its first Citizens' Assembly.[29] The Citizens' Assembly on Social Care brought together a group of forty-seven citizens from across England who

---

ideals and social science principles, Dahl envisioned an innovative mechanism for involving citizens in dealing with public issues. He called it "minipopulus": an assembly of citizens, demographically representative of the larger population, brought together to learn and deliberate on a topic in order to inform public opinion and decision-making' (2017: 1).

[29] The Citizens' Assembly on Social Care was co-commissioned by the Health and Social Care Committee to inform its joint inquiry with the Housing, Communities and Local Government Committee into the long-term funding of adult social care, and in doing so was asked to examine the policy, administration and expenditure of the Department of Health and Social Care and its associated bodies. For more information, see: www.involve .org.uk/our-work/our-projects/practice/how-can-we-find-sustainable-solution-funding-adult-social-care.

convened over two weekends to consider how adult social care should be funded in England. Judging by the comments made by participants, the process of learning from experts with different views and deliberating a range of possible solutions to form a final report for government was very successful – a typical response was 'really enjoyable experience that I'm proud to have been a part of' (Involve, 2018: 24). Based on the lay-members' votes, a final report was written for the government to con-sider, setting out key recommendations with a palpable sense of urgency. Five years later, the UK still has 1.5 million adults with unmet care needs, and one in seven adults aged 65 and over face care costs of at least £100,000 (House of Commons Library, 2023b). While I would argue that this Citizens' Assembly was a highly valuable political exercise, the unfor-tunate point here is that the weight of interest in policy-making was unaltered and little has come of the exercise.

However, this need not be the case. One very particular counter example is that of the Irish Citizens' Assembly (*An Tionól Saoránach*) in July 2016, which illustrates an extraordinary political feat that I suggest has some potential for my argument.

The 2016 Citizens' Assembly was established in Ireland to consider five pressing political questions, including whether to lift the ban on abortion by repealing the Eighth Amendment of the Irish Constitution. The 'Eighth' had been ratified in 1983 and asserted the constitutional personhood of the foetus with a right to life equivalent to that of the mother. This, in effect, secured near complete prohibition in Irish law.[30] Abortion is an issue that had been firmly stuck in political deadlock for decades in a country strongly identified with the Roman Catholic faith and an anti-abortion tradition. As De Londras and Enright describe it,

[30] Exceptions would only be accepted in cases where a mother's life (as opposed to her health) was in serious jeopardy. However, the tragic consequences of such a law were evidenced by the example of Dr Savita Halappanavar. Dr Halappanavar (a dentist originally from India) requested an abortion as she began to suffer a miscarriage. However, she was told by staff at the University Hospital Galway, 'This is a Catholic country and it's the law that they cannot abort when the foetus is live'. Dr Halappanavar miscarried and consequently died of septicaemia a week after entering hospital (BBC, 2012). See De Londras and Enright (2018) for more examples and an in-depth discussion.

the 'Eighth' was a product of 'political turbulence, religious domination and conservative lobbying' (2018: 9), and under it both doctors and pregnant women seeking medical help in life-threatening conditions had to operate 'in the shadow of a custodial sentence' (Dr Rhonda Mahony cited in De Londras and Enright, 2018: 6).

After decades of civil society campaigning by feminist groups, individual activists and women who shared their experiences publicly, demanding change to the Irish Constitution, the Fine Gael party promised a mini-public in 2015 in the run-up to the 2016 General Election, which saw a Fine Gale Independent coalition come to power and with it a Citizens' Assembly on the Eighth.

Without participation by politicians, the Irish Citizens' Assembly comprised a hundred people including a chairperson chosen by the government (in this case a Supreme Court Judge) and ninety-nine citizens (plus ninety-nine substitutes) randomly selected through the electoral register by an independent polling company according to gender, age, location and social class (*An Tionól Saoránach*/Citizens' Assembly, 2017: 24). Administration was provided by the Irish Civil Service and the costs were met by the government and set at €2 million (Department of An Taoiseach, 2019). The media were asked not to photograph members, and no member was paid (although expenses such as travel and child-care were met).

Proceedings began in October 2016, and in its first of many meetings (held in a large Dublin hotel, the Malahide) the Citizens' Assembly focused on considering balanced evidence presented by competing experts with sometimes diametrically different views, submissions from the public, interest groups and institutions (ranging from the Pro-Life Campaign and the Irish Catholic Bishops' Conference to Amnesty International Ireland and the Union of Students in Ireland), having been collected through a nationally advertised website. The participants conducted numerous Q&A sessions with the experts as well as debates and round-table discussions with each other moderated by trained personnel.[31] Finally, proceedings concluded a year and a half later, with a

---

[31] See Parsons (2019) for organisational details such as how independent experts were selected and how the submissions of the general public were provided.

series of secret ballots that returned an overwhelming majority that the Eighth Amendment to the Irish Constitution should be repealed (79 per cent) and a majority vote (51 per cent) on a more technical issue, that rather than introducing new rules in the Constitution, the government should be free to regulate the termination of pregnancy by law (the 36th Amendment).

Crucially, the proceedings were livestreamed for the general public, so that everyone was able to go on a journey of consideration with the citizen members. Finally, the Citizens' Assembly produced a substantial report, including the voting results, to which the government was *legally required to respond* in Parliament (*Oireachtas*). In this sense, the state was fundamentally harnessed to the mechanism of lay-deliberation. The upshot was a national referendum on the Eighth Amendment of the Irish Constitution on 25 May 2018 when the people of Ireland[32] voted by 66.4 per cent to 33.6 per cent to repeal the Eighth Amendment, and brought an end to decades of political impasse that had proved impervious to the usual rounds of competitive democratic elections. As McKee (2018) explains, in 2017 an Irish Times/Ipsos MRBO Poll suggested that only 23 per cent of the public were in favour of legalising abortion in all circumstances in Ireland. However, after engaging with the public deliberations of the 2016 Citizens' Assembly which also acted as the basis of the referendum campaign, the final vote was astonishingly similar to that of the Citizens' Assembly, which had itself voted with a 64 per cent majority in favour of repealing the Eight Amendment and legalising abortion in Ireland. As one Citizen Assembly participant, Fionnuala Geraghty, was recorded as saying:

> I felt relieved the rest of the country listened to the same amount of facts that we heard. I do not think anybody went into this celebrating abortion being anything other than a tragic necessity at times. We were not guided by emotion. We were guided by facts and by experts. It seems that somehow that got into the national consciousness. (McKee, 2018)

It is hard to over-emphasise this extraordinary political achievement whereby religious norms and moral liability were superseded by

---

[32] The turnout was 64.1 per cent (BBC, 2018b).

individual engagement with personal stories, with facts and competing evidence first and foremost. The power of the Citizens' Assembly model lies in its simple but rare deliberative features. As Fishkin argues so convincingly, while the necessary conditions and resources for a well-functioning citizens' assembly are no more than common sense, 'those conditions are, nevertheless, far removed from the routines of everyday life for most citizens most of the time' (2018: 150). It is my view that in this example, we start to see something of what is needed to meet Dunn's call for a 'democratization of prudence' as the most pressing require-ment of modern politics.[33]

The potential for insights into the structural nature of our social connections cannot be captured by the narrow focus of technocrats, professional experts and industry. The opportunity to 'think' in the Arendtian sense under the conditions of a citizens' assembly provides the opportunity to move beyond the limits of a liability-based framework that only requires the identification of agents of fault – the mis-advertisers and the malpractitioners. The policy space must rather be opened up to speculation on the wider structural complexities of com-munal life and understanding this exercise as central to the public interest. The explicit attempt to alter the background conditions of structural injustice is the root to its solution rather than the pursuit of traceable blame to individual acts of wrongdoing. This potential for structural change lies not in duty, nor office nor the charge of liability but rather in the dynamic of ordinary people being exposed to a plurality of situated knowledge and considering competing evidence while oper-ating in conditions practically conducive to the development of political prudence, speculation on macro-level structural dynamics and, crucially, witnessed by many others who join them in their journey of speculation.[34]

That said, the mini-public has its limits and it is these limits, I shall argue, that lead us to a more structured state context such as the public body.

---

[33] See Mounk (2019) for a similar view in the American case.
[34] See, e.g., Habermas (1990) and Rawls (1971). Also see Browne (2013a).

While a well-functioning mini-public can be highly effective as in the case of the 2016 Irish Citizens' Assembly, it is an extremely complicated and expensive model of governance. Mini-publics tend to take a long time to reach a conclusion (often a year or more) and require the sustained efforts of highly trained facilitators as well as a long commitment from selected members. Participants of the mini-public are required to commit to mutually respectful behaviour and consider each other political equals in order for the deliberative process to be successful. However, Shapiro (1999) argues, mini-publics serve to distract from the real power dynamics of politics. His argument is that the powerful political actors are not interested or motivated to enter into respectful and equally weighted political dialogue but instead will operate through other channels such as regulatory capture, which I discussed earlier. Others warn against the idea that mini-publics ought to be detached from the wider public as decision-making entities – rule by the minority: Chambers (2009) and Lafont (2015). However, mini-publics are not usually given the power to directly engage in policy-making, and they are primarily situated outside governance mechanisms with no substantive links to state-level policy decision-making.

With these criticisms in mind, I suggest that some of the positive features of a mini-public, those that speak to Dunn's political prudence, be inserted within a regulatory body model in some ways closer to Warnock's original idea of the HFEA before it was subjected to regulatory gifting as demonstrated in Chapter 3.

THE PUBLIC BODY

While the vast public body landscape[35] has become somewhat unpopular in the public imaginary as a form of technocracy, and for politicians, an association with burdensome regulation, as I have discussed, the status quo need not be so. I argue we need to think again and, in doing so, build an understanding of the public interest that goes well beyond

---

[35] Recall that in the UK the regulatory public body landscape commands a budget of £220 billion with over 300,000 staff.

concerns of liability to those of the structural and institute it through elements of lay-governance directly within state machinery.

The primary feature of this new approach would be to hold the weight of governance towards the public interest largely through a substantial pluralistic lay-presence which would include experiences and perspectives on structural injustices within the policy design process. However, there are several other important aspects. Some of these would continue to include several current public body features:

- The lay-centric public body would be a government-funded recommendatory body, operating at arm's length to elected ministers.
- Whilst ultimately subservient to democratically elected representatives, the public body would ensure that its governance recommendations would be sufficiently independent of political steering.
- Nevertheless, the public body recommendations would be directly harnessed to the state's functions with a requirement of government consideration; this aspect ensures direct lay-member involvement in the formulation of policy recommendations to government. As with the HFEA, Government is then required to consider the recommendations submitted to it from the public body.[36] The recommendations cannot be side-stepped as is the case with the current arrangement of mini-publics in the UK.
- The lay-centric public body would have substantial regulatory powers on settled specified remits concerning the governance of transformative technologies.
- The public body would meet routinely and regularly and lay-members would be included for between 1 and 3 years depending on their ability to continue.
- Members would be remunerated for their work.

Central to the function of the new public body model would be a range of mini-public features:

- Between fifteen and twenty members would be chosen according to a mix of relevant professionals and lay-members with industry and other

---

[36] Recall this was one of the most important features of the 2016 Irish Citizen's Assembly.

experts as non-voting members with the duty to explain and interpret the underlying technicalities of any particular question or concern.

- Lay members would be selected, much like jury service, based on a range of demographic census data (age, ethnicity, gender, geography, socio-economic status, etc.) not because they are expected to perform any particular acts of representation but to ensure the key feature of a plurality of lay-presence.

- The Chair would be selected from within the public body membership and the role rotate between members on a regular basis.

- Members would be exposed to different technical opinions and recommendations and would be given a basic introduction to the public body format designed to encourage innovative thinking and openness to new, sometimes complex and technical ideas. Members would be given a purposefully speculative brief on the structural dynamics of transformative tech, in addition to questions of liability.

- The explicit commitment to speculate beyond the usual liability-based concerns of regulators is key. This means asking lay-members to purposefully reflect on the relationship between a negative conception of the public interest (as discussed in Chapter 4) and industry private interests in the wider social context of technologies such as AI or Repro-tech and their own perspectives and experiences. If such a rotational lay element were built into the otherwise expert deliberations of AI or Repro-tech policy-making, it is much more likely that the macro-level structural dynamics of AI will be addressed in addition to concerns more squarely captured by the standard liability-based approach. Here we might think back to Benjamin's call to think against the emergence of a digital caste system; can we, for example, get the Mechanical Turk to operate differently? Or the public body might consider whether automation of certain labour market tasks is desirable on structural grounds and if not, consider the incentives the state has in its powers such as increasing taxes to compensate for increasing human labour loss to AI. Or there might be a question of licensing genetic technologies on the basis of public access. Whatever the broader structural questions, it is highly likely that members of the public would draw substantially different conclusions from those of industry experts. Such a model enables the public interest to be thought of, not in terms of a presumed inevitability of technological progress and solutionism but rather in terms of the sorts of societies we might actually want to live in. Structural responsibility as a particular strand of collective political responsibility[37] is achieved by a purposeful commitment to speculate on an infinitely wider spectrum of tech-related considerations than market function risk reduction (the current orientation of the 'public interest') along the lines discussed in Chapters 2 and 3. By virtue of the lay-centric focus, the process of keeping a check on private interests is to some degree built directly into the lay-governance of industry through this new public body model.

- To serve on a public body such as this is a discretionary imperfect form of political (structural) responsibility and the focus of speculation, while bounded by the general remit of the particular body (in this case the focus has been either AI or Repro-tech), will change regularly. This is a very different prospect to the form of liability required by Nussbaum or others who argue that political responsibility for particular structural injustice must be grounded in liability as discussed in Chapter 1.
- Members' discussions would be audio or visually live streamed, much like parliament, and votes (where relevant) would be anonymous. This is to ensure the consistency of the format and accountability.

The inclusion of lay members in the routine, systematic and direct functions of policy-making does not of course create a perfect representation of the plurality of experience, however, a specially designed lay-centric format produces a process that will not otherwise occur 'naturally' in technocratic governance. Rather, this lay element is the most practical link to a more diverse set of public concerns and interests that, if built into the deliberations of law and policy-making, will at least ensure that the public interest is more likely to override that of private or industry interest, and rebalance some of the background conditions of structural injustice. I suggest that this approach is a simple manifestation

---

[37] Recall from Chapter 1, that structural responsibility is focused on the exercise of speculating and addressing the background conditions of structural injustice, and is a subset of the wider concept of political responsibility that pertains to both liability-based and structural injustice.

of Dunn's prudent politics but that it also has the ability to affect significant structural change. Because structural injustice is not traceable to agents of fault, as such, I have suggested that the most effective way of enacting structural responsibility is to focus on the background conditions of structural injustice where we see the weighting of public to private interests. In simple terms, creating fora within our governance systems that, with elements of the mini-public model, require that lay members and specialists speculate on the nature of structural actions is a great deal more promising that the current policy-making practices in which a proxy for public interest is minimising direct harms grounded in liability for the purposes of facilitating trust in market products and services.

As I have suggested, almost by design, current democratic engagement with the public such as state consultation processes do not attract the views of those most disconnected from complex technical policy-making circles or those most likely to suffer from structural injustice. It is not that the lay-centric public body can provide a simple solution to each structural injustice but it does ensure engagement with a plurality of situated knowledge where members of the public are able to bring their concerns, experiences and stories to make sense of their relational connections to the background conditions of structural injustice, directly within the governance process.

This is not to deny the importance of experts or industry whose knowledge is so clearly needed for such complex topics as the governance of AI and Repro-tech, but I argue that to bring a range of lay voices into the mix of discussions around the fundamental questions of AI's and Repro-tech's structural impact on society would nevertheless be a fundamental change to the way in which the state currently formulates policy and regulation. It is clear that even with a strong lay element, a new public body for AI governance or a new HFEA model will not provide a simple mapping of 'solution to problem'. Indeed, the opacity of structural injustice does not permit such a traceable route, but without a lay-centric orientation in our policy deliberations, governance will inevitably continue to focus predominately on market-enhancing liability mitigation rather than broader speculative structural accounts of AI, Repro-tech and their influences on society. This sort of structural perspective ought not to be under-estimated, as is currently the case within the mechanisms of contemporary regulation and policy-making in AI and Repro-tech.

AN UNCOMFORTABLE POLITICAL REALISATION

As I set out in Chapter 1, Young's work on structural injustice leads us to an uncomfortable political realisation. The usual tools deployed for addressing injustice – the tracing of liability in contextual moral or legal terms – are not useful for alleviating structural injustice, which is much more complex and amorphous in shape. Even though no simple solution is apparent, I have argued that one useful approach is to focus specifically on the question of whose interests are at play in the construction of the state's laws and regulations which are deeply imbricated in the background conditions of structural injustice. I have argued that reshaping the regulatory public body landscape to incorporate a lay-centric element can only improve current efforts to address structural injustice. Contra Young's suspicions, I have argued here that there is the possibility for state machinery to become more directly and purposefully attuned to such a task.

CHAPTER 5

Conclusion

Technology Is the Answer, What Was the Question?

I hope that the AI Community and governments that fund it will stop saying 'But really bad things cannot happen because the grown-ups will take care of it.'
AI scientist, Stuart Russell (2022)

S O MANY TALKS, PODCASTS, ARTICLES AND BOOKS ON THE pros and cons of AI or Repro-tech end with a plea for 'someone' – whether it be 'you', the 'younger generation' or the 'grown-ups' – to take responsibility for protecting humanity from the spectre of tech-generated harm. What this 'someone' is supposed to do is always opaque. There is never a blueprint offered of what should happen next but there is always a description of how bad things will be if responsibility is not taken up quickly enough.

In this book, I have tried to think about what might practically enable such a responsibility. In order to do this, I chose to begin in a rather odd place – the work of Iris Marion Young. I revisited Young's political theory, not because she worked on the governance of technology but because her work on structural injustice enabled me to explore the limits of a politics grounded in liability, which is where most of the thinking on tech governance currently resides.

While technologies such as AI and Repro-tech have a range of aspects that perhaps, for now at least, can be kept in check for the most part by the liability-based approaches of the law and institutional cultures, I think it is already clear that these technologies are becoming pro-foundly intertwined in the wider structural dynamics of human society. For this reason, it is vital that we speculate politically about what struc-tural challenges may come and how to address them.

Young's untimely death left us with an unfinished manuscript on structural injustice, which, despite its gaps and contradictions, is nevertheless extraordinary. As I have explained, my intention has not been to fill those gaps or second-guess Young's full intentions but rather to take the ideas that I found most inspirational in her work, and to reshape and build on them for a new context.

For me, the question of 'who is responsible where nobody is liable' for the structural dynamics of technological development is a profoundly important political question, and has acted as the central axis of this book. Because structural dynamics are not readily traceable like fault-based causes of injustice, I have tried to grapple with the implications of this conceptual point for the political context and considered the question of what we might practically do to address the background conditions of structural injustice, where, I have argued, we find our best opportunity for significant structural change.

Young most likely would have said that the answer lay in social movements where people, unbounded by national borders or citizenship status, come together to learn and act collectively to effect change at the national and international levels through protest, pressure tactics and, often, impressive social science or scientific research aided by experts. This is vital political work. However, at this stage of humanity's relationship with AI and Repro-tech, significant civil society movements are yet to emerge. Even as they might, these are political entities that operate on the fringes of the mechanisms of state power and are, in virtue of their relationship to the state, limited in their approach to large-scale change, albeit essential in shining a light on the failings of governments. My argument in this book is for fundamental change to the state machinery itself, on the basis that the coordinating power of the state is needed for the scale of structural change required. As governance currently stands, Young was right that the state is a disappointing prospect for addressing structural injustice. However, I have argued that it need not be so.

Think back to where I started this book with the words of futurist Amy Webb:

> There is no one big solution – no big switch that can be flipped – but we can make a thousand incremental changes through the decisions that we

are making everyday. I know that everyone is exhausted, but it's what we have to do!

My aim has been to think how we might bring this sort of necessary action into the sphere of democratic governance for the sorts of tech-generated harms that emanate from beyond liability-based causes.

As the historian Snyder (2023) explains, we should think of democracy not as a noun – as something we 'have' or something in 'decline' – but, rather, as a verb. Democracy is something we must actively 'do'. Political institutions like the ones in my home country, the UK, have become, as Runciman (2019) describes '[m]isfiring democracies – they are old, they are tired, they are stuck in their ways'. More to the point, as Young rightly pointed out, states are currently far too bound up in the interests that serve the background conditions of structural injustice. Technological solutions to humanity's ills have so often suffered from what Haraway described as the 'God trick' (1988: 584), a perspective seemingly from 'above, from nowhere' (589) grounded in science that is presented as neutral and universally objective, thereby dominating over all other perspectives. In fact, technology and its designers could only ever provide a partial view of the world, and never one that is neutral.[1] Our political institutions, however, tend not to take account of such limitations, instead bestowing a great deal of power and influence on those with expert status. Driven by the fear of stunting innovation and investment, the UK Government has exemplified what it means to 'gift deregulation' to the tech sector under the guise of public interest grounded in oversimplified narratives of economic growth and national supremacy in the perpetual global technology race. So much of the decision-making on the use and design of technology is removed from the public eye and, therefore, from its consciousness. In what sense, could the average person, 'Jane, Joe or José',[2] ever make a difference to the juggernaut of AI or genetic technologies? The inclusion of situated knowledge to policy-making on these technologies perhaps, on the

[1] The anti-tech fundamentalists who reject technological innovation whatever the evidence (e.g., anti-vaxxers) also belong to a belief system that is closed to other perspectives.

[2] See 'Introduction', where I begin with Wajahat Ali's interview with futurist Amy Webb about how we might address tech-generated harm to human society.

face of it, seems not only impracticable but woefully insignificant. I have argued, however, that it is neither.

This brings me to what has been the hardest part of my argument to articulate (although somewhat ironically, it is, in some ways, a simple point): why lay-centric politics is important for addressing structural injustices that are untraceable to agents of fault.

The argument is not that there is some kind of teleological conception of the public interest that if achieved will cancel out structural injustice. The argument is not that all, nor nearly all, characteristics and perspectives can be represented in decision-making contexts through lay inclusion. The argument is not that a specific kind of situated knowledge will change the design of a given technological innovation. Rather, the argument is that by creating the right conditions for structural speculation, the balance of private and public interests, which operate in the background conditions of structural injustice, will most likely be altered when ordinary people bring their own experiences, concerns, fears, perspectives and stories to the details of a complex political decision in real time. Politics is always messy and unpredictable. However, such a direction is what I think Dunn was hoping for when he called for 'the democratization of prudence' (1990: 214).

Of course, it may seem that the closest we have come to the purest form of democratisation, direct democracy, are our exchanges with each other through social media. Hundreds of millions of people have the opportunity to converse and form opinions on the political issues of the day, any time of the day, through X (formerly known as Twitter), Facebook and similar platforms. This, however, is not democracy but rather, more of an echo chamber for like-minded beliefs or, worse still, exposure to conspiracy and fakery. What is missing are the conditions of serious deliberation, ordinary people who have some sense of 'ordinary injustices' faced with expert evidence and political interpretations as to their effects on the world. This sort of exercise requires a specific format. Certainly, there is an emerging interest in the idea of using mini-publics and citizens assemblies to address seemingly intractable political problems. Yet, like civil society movements, these models of lay-centric deliberation often tend to operate outside the sort of state machinery that has sufficient coordinating power to effect change at the structural level.

I have advocated for the introduction of lay-centric membership of the inner workings of states' substantial regulatory machinery as a means to addressing the background conditions of structural injustice. Lay-centric thinking does not, in and of itself, give us a clear causal picture of the structural dynamics of technological impact on society. It is however, a much more likely way for societies to rebalance the structural conditions of technological development, its design, use, access and sometimes its rejection, in favour of the public interest. So rather than load up future generations with the responsibility to correct our mistakes or defer to the illusive 'grown-ups', this is something we can do, and I argue, must do, now, to steer the future to a place where we know the question to which technology is the answer.

Legal, Moral and Structural Actions, Injustices and Responsibilities

Table A.1 Young's different forms of responsibility for 'ordinary injustices'

Forms of duty or responsibility	Legal responsibility (liability-based duty to follow the law)	Moral responsibility (liability-based duty to existing moral norms)	Political responsibility (discretionary, not a duty)
Model of responsibility	Liability model		Social connection model
Forms of injustice	Legal injustices (traceable acts of illegality)	Moral injustices (traceable transgressions of societal norms)	Structural injustices (the untraceable consequences of the actions of masses of individuals, groups and institutions)
Perpetrators of injustice	Any individual, group or institution	Any individual, group or institution	Any individual, group or institution
Injustice grounded in…	Wrongdoing/fault/guilt/liability (direct or indirect)	Wrongdoing/fault/guilt/liability (direct or indirect)	Social connection to background conditions of structural injustice
Responsibility to address injustice grounded in …	Wrongdoing/fault/guilt/liability (direct or indirect); regulated by state or other legal institutions	Wrongdoing/fault/guilt/liability (direct or indirect); regulated by societal norms and moral institutions	Social connection to background conditions of structural injustice acted on with discretion at the level of civil society
Direction of responsibility focus	Backward-facing	Backward-facing	Forward-facing (with an eye to the past)

Table A.2 Browne's different forms of political responsibility for 'ordinary injustices'

Political responsibility			
Subsets of political responsibility:	Legal responsibility (liability/fault-based duty to abide by the law)	Moral responsibility (liability/fault-based duty to abide by existing moral norms)	Discretionary structural responsibility (structural speculation on the untraceable structural actions that contribute to the background conditions of structural injustice)
Forms of injustice	Legal injustices (traceable acts of illegality)	Moral injustices (traceable transgressions of societal norms)	Structural injustices (consequence of mass untraceable structural actions)
Perpetrators of injustice	Any individual, group or institution which breaks the law	Any individual, group or institution which transgresses social norms	Any individual, group or institution contributing to the background conditions of macro-level structural injustice
Injustice grounded in …	Legal acts of wrongdoing/fault/guilt/liability (direct or indirect)	Moral acts of wrongdoing/fault/guilt/liability (direct or indirect)	Structural actions The 'legitimate' pursuits of private interest in a given time and space (actions, expressed beliefs and habits) that we might speculate contribute to the background conditions of a particular structural injustice
Responsibility to address injustice grounded in…	Abidance by the law to address legal acts of wrongdoing/fault/guilt/liability (direct or indirect); regulated by state, legal and other institutions/civil society/individuals	Abidance by moral codes to address moral acts of wrongdoing/fault/guilt/liability (direct or indirect); regulated by societal norms and moral institutions/civil society/individuals	Structural actions potentially connected to structural injustice – considered through imperfect political (structural) responsibility enacted by state, legal and other institutions/civil society/individuals
Direction of responsibility focus	Backward-facing (with an eye to the future)	Backward-facing (with an eye to the future)	Forward-facing (with an eye to the past)
Potential direction of transition between different forms of injustice depending on new epistemologies	(→) (possibility that traceable legal injustice reverts back to traceable moral injustice)	← (when traceable moral injustice becomes traceable legal injustice)	← (when untraceable structural injustice becomes traceable moral injustice)

161

References

13D (2018) For Emmanuel Macron, AI is more than a technological revolution. It is a political revolution of hope in an increasingly dystopian future (12 May). Available at https://latest.13d.com/emmanuel-macron-ai-political-revolution-dystopian-ai-france-gdpr-4e60d4090c54 (Accessed 2 September 2023).

Achen, Christopher and Bartels, Larry (2017) *Democracy for Realists: Why Elections Do Not Produce Responsive Government.* Princeton, NJ: Princeton University Press.

Agar, Nicholas (2004) *Liberal Eugenics: In Defense of Human Enhancement.* Oxford: Wiley-Blackwell.

Ahmed, Sara (2012) *On Being Included: Racism and Diversity in Institutional Life.* Durham, NC and London: Duke University Press.

(2013) *The Cultural Politics of Emotion.* London: Routledge.

(2017) *Living a Feminist Life.* London: Duke University Press.

(2021) *Complaint.* London: Duke University Press.

Ahuja, Abhimanyu S. (2019) The impact of artificial intelligence in medicine on the future role of the physician. *PeerJ* 7: e7702. https://doi.org/10.7717/peerj.7702.

Allen, Danielle S. (2016) Toward a connected society. In E. Lewis and N. Cantor (eds.), *Our Compelling Interests: The Value of Diversity for Democracy and a Prosperous Society.* Princeton, NJ: Princeton University Press, pp. 71–105.

Allen, Danielle (2023) *Justice by Democracy.* Chicago, IL: University of Chicago Press.

Aljawoan, Fatimah, Hunt, Linda and Gordon, Uma (2012) Prediction of ovarian hyperstimulation syndrome in coasted patients in an IVF/ICSI program. *Journal of Human Reproductive Science* 5(1): 32–36.

Amoore, Louise (2019) Doubt and the algorithm: On the partial accounts of machine learning. *Theory, Culture & Society* 36(6): 147–169. Available at https://dro.dur.ac.uk/26913/ (Accessed 2 September 2023).

Anderson, Porter (2023) European publishers welcome Parliament's 'AI Act' draft approval. Publishing Perspectives (14 June). Available at

https://publishingperspectives.com/2023/06/european-publishers-wel
come-parliaments-ai-act-draft-approval/ (Accessed 2 September 2023).

Aragon, Corwin and Jaggar, Alison (2018) Agency, complicity, and the responsibility to resist structural injustice. *Journal of Social Philosophy* 49(3) Special
issue: Frontiers of Responsibility for Global Justice: 439–460.

Arendt, Hannah (1968) *Men in Dark Times.* New York: Harcourt Brace & Company.

(1971) Thinking and moral considerations: A lecture. *Social Research* 38:
430–448.

(1994 [1945]) Organized guilt and universal responsibility. In Jerome Kohn
(ed.), *Essays in Understanding 1930–1954: Formation, Exile, and Totalitarianism.*
New York: Schocken Books, pp. 121–132.

(1998 [1958]) *The Human Condition.* Chicago, IL: University of Chicago Press.

(2003 [1968]) Collective responsibility. In Jerome Kohn (ed.), *Responsibility and
Judgement.* New York: Schocken Books, pp. 147–158.

(2003 [1964]) Personal responsibility under dictatorship. In Jerome Kohn
(ed.), *Responsibility and Judgement.* New York: Schocken Books, pp. 17–48.

Arendt, Hanna (2005 [1954]) Socrates. In Jerome Kohn (ed.), *The Promise of
Politics.* New York: Schocken Books, pp. 5–39.

(2005 [1963]) Introduction into politics. In Jerome Kohn (ed.), *The Promise of
Politics.* New York: Schocken Books, pp. 93–200.

(2006 [1963]) *Eichmann in Jerusalem: A Report on the Banality of Evil.* London:
Penguin Classics.

Arntz, Melanie, Gregory, Terry and Zierahn, Ulrich (2016) The risk of automation
for jobs in OECD countries: A comparative analysis, OECD Social, Employment
and Migration Working Papers, No. 189, Paris: OECD Publishing.

Atenasio, David (2019) Blameless participation as structural injustice. *Social
Theory and Practice* 45(2): 149–177.

Attenborough, David (2019) David Attenborough speaks in parliament about
climate change. Online video clip. Available at www.youtube.com/watch?v=
rv3DPaMaS2g (Accessed 2 September 2023).

Austrian World Summit (2021) Greta Thunberg speech. Online video clip. Available
at www.youtube.com/watch?v=m6eQwAi2U18 (Accessed 2 September 2023).

Bächtiger, André, Dryzek, John S., Mansbridge, Jane J. and Warren, Mark (eds.)
(2018) *The Oxford Handbook of Deliberative Democracy.* Oxford: Oxford
University Press

Baldwin, Kylie (2017) 'I suppose I think to myself, that's the best way to be a
mother': How ideologies of parenthood shape women's use for social egg
freezing technology. *Sociological Research Online* 22(2): 2.

Baldwin, Kylie, Culley, Lorraine, Hudson, Nicky and Mitchell, Helene (2018)
Running out of time: Exploring women's motivations for social egg freezing.

Journal of Psychosomatic Obstetrics & Gynecology 40(2): 166–173 (Accessed 2 September 2023).

Bao, Min, Cornwall-Scoones, Jake, Sanchez-Vasquez, E. et al. (2022) Stem cell-derived synthetic embryos self-assemble by exploiting cadherin codes and cortical tension. *Nature Cell Biology* 24: 1341–1349.

Barry, Brian (1964) The public interest. *Proceedings of the Aristotelian Society* 38 (supp.): 1–18.

(1965) *Political Argument*. New York: Humanities Press.

Barry, C. and Ferracioli, L. (2013) Young on responsibility and structural injustice. *Criminal Justice Ethics* 32(3): 247–257.

Barry, Christian and Macdonald, Kate (2016) How should we conceive of individual consumer responsibility to address labour injustices? In Yossi Dahan, Hanna Lerner and Faina Milman-Sivan (eds.), *Global Justice and International Labour*. Cambridge: Cambridge University Press, pp. 92–118.

Barratt, James (2013) *Our Final Invention: Artificial Intelligence and the End of the Human Era*. London: Quercus Press.

BBC (2012) Woman dies after abortion request 'refused' at Galway hospital (14 November). Available at www.bbc.co.uk/news/uk-northern-ireland-20321741 (Accessed 30 August 2022).

(2014) Stephen Hawking warns artificial intelligence could end mankind (2 December). Available at www.bbc.co.uk/news/technology-30290540 (Accessed 2 September 2023).

(2017) Donald Trump tells banks he will give laws a 'haircut' (4 April). Available at www.bbc.co.uk/news/business-39498664 (Accessed 30 August 2022).

(2018a) Bank of England chief economist warns on AI jobs threat (20 August). Available at www.bbc.co.uk/news/business-45240758 (Accessed 2 September 2023).

(2018b) Irish abortion referendum: Ireland overturns abortion ban (26 May). Available at www.bbc.co.uk/news/world-europe-44256152 (Accessed 2 September 2023).

(2020) Uber's self-driving operator charged over fatal crash (16 September). Available at www.bbc.co.uk/news/technology-54175359 (Accessed 2 September 2023).

Beardsworth, R. (2015). From moral to political responsibility in a globalized age. *Ethics & International Affairs* 29(1): 71–92.

Beck, Valentin (2020) Two forms of responsibility: Reassessing Young on structural injustice. *Critical Review of International Social and Political Philosophy* 26 (6): 918–941.

Becker, Gary (1983) A theory of competition among pressure groups for political influence. *Quarterly Journal of Economics* 98: 371–400.

Becker, G. (1991) *A Treatise on The Family.* Cambridge, MA: Harvard University Press.

Beebee, Helen, Hitchcock, Christopher and Menzies, Peter (eds.) (2012) *The Oxford Handbook of Causation.* Oxford: Oxford University Press.

Bendix, Aria (2014) Forget babies, The future of sex is recreational (17 November 17). Available at www.bustle.com/articles/49573-all-western-babies-will-be-born-via-ivf-by-2050-predicts-pill-inventor-professor-carl-djerassi (Accessed 5 September 2023).

Benhabib, Seyla (1992) *Situating the Self: Gender, Community and Postmodernism in Contemporary Ethics.* London: Routledge.

(1993) Feminist theory and Hannah Arendt's theory of the public space. *History of the Human Sciences* 6(2): 97–114.

(1999) The personal is not the political. *Boston Review* (1 October). Available at https://bostonreview.net/articles/seyla-benhabib-personal-not-political/ (Accessed 2 September 2023).

(2018) *Exile, Statelessness and Migration: Playing chess with history from Hannah Arendt to Isaiah Berlin.* Princeton, NJ: Princeton University Press.

(2021) Gender and emigré political thought: Hannah Arendt and Judith Shklar. In Jude Browne (ed.), *Why Gender?* Cambridge: Cambridge University Press, pp. 267–288.

Benjamin, Ruha (2019) *Race after Technology: Abolitionist Tools for the New Jim Code.* Cambridge: Polity Press.

Bentham, Jeremy (2007 [1780]) *Introduction to the Principles of Morals and Legislation.* Mineola, NY: Dover.

Bernstein, Marver H. (1955) *Regulating Business by Independent Commission.* Princeton, NJ: Princeton University Press.

Bickerton, Christopher and Accetti, Carlo Invernizzi (2017) Populism and technocracy: Opposites or complements? *Critical Review of International Social and Political Philosophy* 20(2): 186–206.

Bivens, Josh and Lawrence, Mishel (2021) The failure of automation and skill gaps to explain wage suppression or wage inequality. Economic Policy Institute (20 May). Available at www.epi.org/unequalpower/publications/automation-myth/ (Accessed 2 September 2023).

Bonefeld W (2017) *The Strong State and the Free Economy.* London: Rowman & Littlefield.

Bostrom, Nick (2014) *Superintelligence: Paths, Dangers and Strategies.* Oxford: Oxford University Press.

Bostrom, Nick and Sandberg, Anders (2017) The wisdom of nature: An evolutionary heuristic for human enhancement. In D. Ho (ed.),

Philosophical Issues in Pharmaceutics. Philosophy and Medicine. Dordrecht: Springer, vol. 122, pp. 189–219.

Bourdieu, Pierre (1984) *Distinction: A Social Critique of the Judgement of Taste,* trans. Richard Nice. Cambridge, MA: Harvard University Press.

Braidotti, Rosi (2013) *The Posthuman.* Cambridge: Polity Press.

(2022) Posthuman feminism and gender methodology. In Jude Browne (ed.), *Why Gender?* Cambridge: Cambridge University Press, pp. 101–125.

Branson, Richard. 2015. Sandberg and Branson defend Facebook's egg-freezing policy. 24th April interview with Emily Chang. New York: Bloomberg. Available at http://time.com/3835233/sheryl-sandberg-explains-why-facebook-covers-egg-freezing/ (Accessed 2 September 2023).

Brennan, T., Dieterich, W. and Ehret B. (2009) Evaluating the predictive validity of the COMPAS risk and needs assessment system. *Criminal Justice and Behavior* 36(1): 21–40.

Brooke, Ackerly (2018) *Just Responsibility.* Oxford: Oxford University Press.

Brown, Wendy (2019) *In the Ruins of Neoliberalism: The Rise of Anti-Democratic Politics in the West.* New York: Columbia University Press.

Browne, Jude (2006) *Sex Inequality and Sex Segregation in the Modern Labour Market.* Bristol: The Policy Press.

(ed.) (2013a) *Dialogue, Politics and Gender.* Cambridge: Cambridge University Press.

(2013b) O'Neill and the political turn against human rights. *International Journal of Politics, Culture and Society* 26(4): 291–304.

(2013c) The default model: Gender equality and structural constraint. *Politics & Gender* 9(02): 152–173.

(2014) The critical mass marker approach: Female quotas and social justice. *Political Studies* 62(4): 862–877.

(2020) 'The Regulatory Gift: Politics, Regulation and Governance', *Regulation and Governance* 14(2): 165–388.

Browne, Jude and McKeown, Maeve (eds.) (2024) *What Is Structural Injustice?* Oxford: Oxford University Press, pp. 1–11.

Browne, Jude and Stears Marc (2005) Capabilities, resources, and systematic injustice: A case of gender inequality. *Politics, Philosophy and Economics* 4(3): 355–373.

Browne, Jude, Cave, Stephen, Drage, Eleanor and McInerney, Kerry (eds.) (2023) *Feminist AI: Critical Perspectives on Algorithms, Data, and Intelligent Machines.* Oxford: Oxford University Press.

Brynjolfsson, Erik (2021) The problem is wages, not jobs. In Daron Acemoglu (ed.), *Redesigning AI, Work, Democracy and Justice in the Age of Automation.* Cambridge, MA: MIT Press, pp. 65–71.

Brynjolfsson, Erik and McAfee, Andrew (2014) *The Second Machine Age: Work, Progress, and Prosperity in a Time of Brilliant Technologies.* New York: W. W. Norton & Company.

Buchanan, Allen, Brock, Dan, Daniels, Norman and Wikler, Daniel (2000) *From Chance to Choice.* Cambridge: Cambridge University Press.

Buolamwini, Joy and Gebru, Timnit (2018) Gender shades: Intersectional accuracy disparities in commercial gender classification. *Proceedings of Machine Learning Research* 81: 1–15.

Butler, Judith (2003) *Giving an Account of Oneself.* New York: Fordham Press.

(2004) *Undoing Gender.* London: Routledge.

(2011) Hannah Arendt's death sentences. *Comparative Literature Studies* 48(3) Special issue, Trials of Trauma: 280–295.

(2018) *Notes Toward a Performative Theory of Assembly.* Cambridge, MA: Harvard University Press.

(2020) quoted in Gessen, Masha (2020) Judith Butler wants us to reshape our rage. *The New Yorker* (9 February). Available at www.newyorker.com/culture/the-new-yorker-interview/judith-butler-wants-us-to-reshape-our-rage.

Cabinet Office (2010) Public Body Review Published. The Right Honourable Lord Maude of Horsham. UK Cabinet Office Press Release. London: HMG. Available at www.gov.uk/government/news/public-body-review-published (Accessed 12 January 2023).

(2015) *Public Bodies 2015.* London: Cabinet Office, the UK Government. Available at www.gov.uk/government/uploads/system/uploads/attachment_data/file/506880/Public_Bodies_2015_Web_9_Mar_2016.pdf (Accessed 15 February 2024).

Cabinet Office (2016) Classification of Public Bodies: Guidance for Departments. Available at https://assets.publishing.service.gov.uk/government/uploads/system/uploads/attachment_data/file/519571/Classification-of-Public_Bodies-Guidance-for-Departments.pdf (Accessed 2 September 2023).

(2017) Open for Business Campaign. Available at www.gov.uk/government/publications/open-for-business-campaign (Accessed 4 April 2024)

(2019a) Public Bodies 2018–19. Available at www.gov.uk/government/publica tions/public-bodies-2018-19-report (Accessed 2 September 2023).

(2019b) Tailored review guidance on public bodies. London: HMG. Available at https://assets.publishing.service.gov.uk/government/uploads/system/uploads/attachment_data/file/802961/Tailored_Review_Guidance_on_public_bodies-May-2019.pdf (Accessed 2 September 2023).

(2021) Declaration on government reform. Available at assets.publishing.service .gov.uk/government/uploads/system/uploads/attachment_data/file/993902/

FINAL_Declaration_on_Government_Reform.pdf, p. 10 (Accessed 2 September 2023).

(2022) Guidance on the Undertaking of Reviews of Public Bodies. Press release, HM Government. Available at www.gov.uk/government/publications/public-bodies-review-programme/guidance-on-the-undertaking-of-reviews-of-public-bodies (Accessed 12 January 2023).

(2023) Guidance; Public bodies. Available at www.gov.uk/guidance/public-bodies-reform#public-bodies-reform-programme-2020-to-2025 (Accessed 2 September 2023).

Caplan, Arthur (2001) What should the rules be? *Time Magazine* (14 January). Available at https://content.time.com/time/magazine/article/0,9171,95244,00 .html (Accessed 2 September 2023).

Carpenter, Daniel (2014) Corrosive capture? The dueling forces of autonomy and industry influence in FDA pharmaceutical regulation. In Daniel Carpenter and David A. Moss (eds.), *Preventing Regulatory Capture: Special Interest Influence and How to Limit It.* Cambridge: University of Cambridge Press, pp. 152–172.

Carpenter, Daniel and Moss, David A. (2014a) Introduction. In Daniel Carpenter and David A. Moss (eds.), *Preventing Regulatory Capture: Special Interest Influence and How to Limit It.* Cambridge: University of Cambridge Press, pp. 1–22.

(2014b) New conceptions of capture – Mechanisms and outcomes. In Daniel Carpenter and David A. Moss (eds.), *Preventing Regulatory Capture: Special Interest Influence and How to Limit It.* Cambridge: University of Cambridge Press, pp 69–70.

Castelvecchi, Davide (2016) Can we open the black box of AI? *Nature* 538(7623): 20–23.

Centre for Data Ethics and Innovation (CDEI) (2020) Review into bias in algorithmic decision-making. Available at https://assets.publishing.service .gov.uk/government/uploads/system/uploads/attachment_data/file/957259/Review_into_bias_in_algorithmic_decision-making.pdf (Accessed 2 September 2023).

(2022) Britainthink; Insight & Strategy. Available at https://assets.publishing .service.gov.uk/government/uploads/system/uploads/attachment_data/file/1177293/Britainthinks_Report_-_CDEI_AI_Governance.pdf (Accessed 2 September 2023).

Chaffey, Dave (2022) Global social media statistics research summary 2022. Smart Insights (22 August). Available at www.smartinsights.com/social-media-marketing/social-media-strategy/new-global-social-media-research (Accessed 2 September 2023).

Chambers, Simone (2009) Rhetoric and the public sphere: Has deliberative democracy abandoned mass democracy? *Political Theory* 37(3): 323–350.

Cheng, Andria (2018) How Adidas plans to bring 3D printing to the masses. *Forbes* (22 May). Available at www.forbes.com/sites/andriacheng/2018/05/22/with-adidas-3d-printing-may-finally-see-its-mass-retail-potential/ (Accessed 2 September 2023).

Christensen, Jørgen Grønnegård (2011) Competing theories of regulatory governance: Reconsidering public interest theory of regulation. In David Levi-Fair (ed.), *Handbook on the Politics of Regulation*. Northampton, MA: Edward Elgar Publishing, pp. 96–112.

Chung, Ryoa (2021) Structural health vulnerability: Health inequalities, structural and epistemic injustice. *Journal of Social Philosophy* 52(2) Special issue: Global Justice and Structural Injustice: 201–216.

CIPHR (2022) Will your job exist in the future? Occupations most at risk from automation. Available at www.ciphr.com/jobs-at-risk-from-automation/ (Accessed 2 September 2023).

Clarke, Desmond (2014) Causation and liability in Tort Law. *Jurisprudence* 5(2): 217–243.

Coco, A. (2023) Exploring the impact of automation bias and complacency on individual criminal responsibility for war crimes. *Journal of International Criminal Justice*, https://academic.oup.com/jicj/advance-article/doi/10.1093/jicj/mqad034/7281035

Collins, Sarah and Garget, Jacqueline (2022) 'Synthetic' embryo with brain and beating heart grown from stem cells by Cambridge scientists. University of Cambridge website (25 August). Available at www.cam.ac.uk/stories/model-embryo-from-stem-cells (Accessed 2 September 2023).

Cooper, Charlie (2020) Boris Johnson turns to polling and 'common sense'. *Politico* (13 May). Available at www.politico.eu/article/boris-johnsons-coronavirus-fudge/ (Accessed 2 September 2023).

Costanza-Chock, Sasha (2023) Design practices: 'Nothing about us without us'. In Jude Browne, Stephen Cave, Eleanor Drage and Kerry McInerney (eds.), *Feminist AI: Critical Perspectives on Algorithms, Data, and Intelligent Machines*. Oxford: Oxford University Press, pp. 370–388.

Courtland, R. (2018) Bias detectives: The researchers striving to make algorithms fair. *Nature* 558: 357–360.

Cowls, Josh and Floridi, Luciano (2018) Prolegomena to a White Paper on an Ethical Framework for a Good AI Society. Available at https://papers.ssrn.com/sol3/papers.cfm?abstract_id=3198732 (Accessed 2 September 2023).

D'Agostino, M. and Durante, M. (2018) Introduction: The governance of algorithms. *Philosophy & Technology* 31: 499–505.

d'Entreves, Maurizio Passerin (2019) Hannah Arendt. *The Stanford Encyclopedia of Philosophy* (Fall edition), Edward N. Zalta (ed.). Available at https://plato.stanford.edu/entries/arendt/ (Accessed 2 September 2023).

Dastin, Jeffrey (2018) Amazon scraps secret AI recruiting tool that showed bias against women. *Reuters* (11 October). Available at www.reuters.com/article/us-amazon-com-jobs-automation-insight/amazon-scraps-secret-ai-recruiting-tool-that-showed-bias-against-women-idUSKCN1MK08G (Accessed 12 January 2023).

Davis, Allison P. (2022) Meghan of Montecito. *The Cut* (29 August). Available at www.thecut.com/article/meghan-markle-profile-interview.html (Accessed 2 September 2023).

Dawes, James (2021) UN fails to agree on 'killer robot' ban as nations pour billions into autonomous weapons research. *The Conversation* (20 December). Available at https://theconversation.com/un-fails-to-agree-on-killer-robot-ban-as-nations-pour-billions-into-autonomous-weapons-research-173616 (Accessed 2 September 2023).

De Londras, Fiona and Enright, Máiréad (2018) *Repealing the 8th.* Bristol: Policy Press, Bristol University Press.

Delacroix, Sylvie (2021) Diachronic interpretability & machine learning systems. *Journal of Cross-disciplinary Research in Computational Law.* https://doi.org/10.2139/ssrn.3728606.

Department for Business, Energy and Industry Strategy (2016) Growth Duty S110 Guidance (December). London: HMG. Available at www.gov.uk/government/uploads/system/uploads/attachment_data/file/574499/growth-duty-statutory-guidance.pdf.

Department for Business, Innovation and Skills (2014) Minister of State for Business and Enterprise. Regulator's Code. London: Better Regulation Delivery Office. HMG. Available at www.gov.uk/government/uploads/system/uploads/attachment_data/file/300126/14-705-regulators-code.pdf.

(2016) Government going further to cut red tape by £10 billion. Press Release (3 March 2016). Department of Business, Innovation and Skills. London: HMG: Available at www.gov.uk/government/news/government-going-further-to-cut-red-tape-by-10-billion (Accessed 22 September 2022).

Department for Digital, Culture, Media and Sport (2021) Government Response to the House of Lords Select Committee on Artificial Intelligence (February). Available at https://assets.publishing.service.gov.uk/government/uploads/system/uploads/attachment_data/file/963696/Government_Response_to_the_HoL_Select_Committee_on_AI_v2.pdf (Accessed 2 September 2023).

Department of an Taoiseach (2019) Written answers. Kildarestreet (26 March). Available at www.kildarestreet.com/wrans/?id=2019-03-26a.175 (Accessed 2 September 2023).

Department of Health (2010) Review of arm's length bodies to cut bureaucracy. Press release, HM Government. Available at https://webarchive.nationalarchives.gov.uk/+/www.dh.gov.uk/en/MediaCentre/Pressreleases/DH_117844 (Accessed 2 September 2023).

(2013) Review of Human Fertilisation & Embryology Authority and Human Tissue Authority. Available at www.gov.uk/government/publications/review-of-human-fertilisation-embryology-authority-and-human-tissue-authority (Accessed 2 September 2023).

(2016). *Response to the HFEA Triennial Review 2016: Nuffield Council of Bioethics.* Department of Health. London: HMG. Available at http://nuffieldbioethics.org/wp-content/uploads/NCOB-response-to-HFEA-triennial-review.pdf.

Department of Health and Social Care (2021) Consultation document: gamete (egg, sperm) and embryo storage limits. Available at www.gov.uk/government/consultations/egg-sperm-and-embryo-storage-limits/consultation-document-gamete-egg-sperm-and-embryo-storage-limits (Accessed 2 September 2023).

Destatis (2023) Labour market: Employment. Available at www.destatis.de/EN/Themes/Labour/Labour-Market/Employment/_node.html (Accessed 3 September 2023).

Devlin, Hannah (2023) Scientist who edited babies' genes says he acted 'too quickly'. *The Guardian* (4 February). Available at www.theguardian.com/science/2023/feb/04/scientist-edited-babies-genes-acted-too-quickly-he-jiankui (Accessed 2 September 2023).

Dillon, Sarah and Craig, Claire (2021) *Storylistening: Narrative Evidence and Public Reasoning.* London: Routledge.

Djerassi, Carl (2014a) Carl Djerassi, father of the contraceptive pill. *Financial Times,* interview with Clive Cookson (21 November). Available at www.ft.com/content/2257f88e-6f77-11e4-b50f-00144feabdc0 (Accessed 2 September 2023).

(2014b) *In Retrospect: From the Pill to the Pen.* London: Imperial College Press.

(2015): Interview: Carl Djerassi, hailed as the father of the pill. *The Independent* (2 February). Available at www.independent.co.uk/news/people/news/carl-djerassi-chemist-hailed-as-the-father-of-the-pill-whose-work-helped-pave-the-way-for-the-sexual-10019100.html (Accessed 2 September 2023).

Dommett, Katherine and Flinders, Matthew (2015). The centre strikes back: Meta-governance, delegation, and the core executive in the United Kingdom, 2010–14. *Public Administration* 93: 1–16.

Dommett, K., Flinders, M., Skelcher, C. and Tonkiss, K. (2014) Did they 'Read Before Burning'? The coalition and Quangos. *The Political Quarterly* 85: 133–142.

Downs, Anthony (1957) An economic theory of political action in a democracy. *Journal of Political Economy* 65(2): 135–150.

Drage, Eleanor and Mackereth, Kerry (2022) Does AI de-bias recruitment? Race, gender and AI's 'eradication of differences between groups'. In Proceedings of the 2022 AAAI/ACM Conference on AI, Ethics, and Society (AIES'22), 1–3 August 2022, Oxford, UK. https://doi.org/10.1145/3514094.353415 (Accessed 2 September 2023).

Dryzek, John (2001) Legitimacy and economy in deliberative democracy. *Political Theory* 29(5): 651–669.

Dunn, John (1990) *Interpreting Political Responsibility.* Cambridge: Polity Press.

(1992) Conclusion. In John Dunn (ed.), *Democracy: The Unfinished Journey 508 BC to AD 1933.* Oxford: Oxford University Press, pp. 239–267.

(2012) Fatalismes contrefaits et manquements réels dans les démocraties contemporaines. In Charles Zarka Yves (ed.), *Démocratie, état critique.* Paris: Armand Colin, pp. 199–213.

Dupré, John (2021) Gender and the end of biological determinism. In Jude Browne (ed.), *Why Gender?* Cambridge: Cambridge University Press, pp. 57–77.

Dworkin, Ronald (2002) *Sovereign Virtue: The Theory and Practice of Equality.* Cambridge, MA: Harvard University Press.

(2001) Legitimacy and economy in deliberative democracy. *Political Theory* 29 (5): 651–669.

Dyer, Silke, Archary, Paversan, Potgieter Liezel et al. (2020) Assisted reproductive technology in Africa: A 5-year trend analysis from the African Network and Registry for ART. *Reproductive Biomedicine Online* 41(4): 604–615.

Dzieza, Josh (2023) AI is a lot of work. The Verge (20 June). Available at www .theverge.com/features/23764584/ai-artificial-intelligence-data-notation-labor-scale-surge-remotasks-openai-chatbots (Accessed 2 September 2023).

Edwards, Lilian (2022) *The EU AI Act proposal.* Ada Lovelace Institute. Available at www.adalovelaceinstitute.org/resource/eu-ai-act-explainer/ (Accessed 4 April 2024).

Eriksen, Andreas (2020) Political values in independent agencies. *Regulation and Governance* 15(3): 785–799.

Escobar, Oliver and Elstub, Stephen (2017) Forms of mini-publics. New Democracy Foundation (8 May). Available at www.newdemocracy.com.au/docs/research notes/2017_May/nDF_RN_20170508_FormsOfMiniPublics.pdf (Accessed 2 September 2023).

Estudillo, Enrique, Jiménez, Adriana, Bustamante-Nieves, Pablo Edson, Palacios-Reyes, Carmen, Velasco, Iván and López-Ornelas, Adolfo (2021) Cryopreservation of gametes and embryos and their molecular changes. *International Journal of Molecular Sciences* 22(19): 10864.

Eubanks, V. (2018) *Automating Inequality: How High-Tech Tools Profile, Police and Punish the Poor.* New York: St Martin's Publishing Group.

EUR-Lex (2021) Artificial Intelligence Act. Available at www.europarl.europa.eu/doceo/document/TA-9-2023-0236_EN.html (Accessed 2 September 2023).

European Commission (2021) Proposal for a Regulation of the European Parliament and of the Council. Laying down harmonised rules on artificial intelligence (Artificial Intelligence Act) and amending certain union legislative acts (21 April). Available at https://eur-lex.europa.eu/legal-content/EN/TXT/?uri=CELEX%3A52021PC0206 (Accessed 2 September 2023).

Eurostat (2023) Fertility Statistics. Published online. Available at https://ec.europa.eu/eurostat/statistics-explained/index.php?title=Fertility_statistics

Evans, Lord Evans of Weardale (2020) Artificial intelligence and public standards: a review by the committee on standards in public life. UK Government. Available at https://assets.publishing.service.gov.uk/government/uploads/system/uploads/attachment_data/file/868284/Web_Version_AI_and_Public_Standards.PDF (Accessed 2 September 2023).

Fauser, Bart (2019) Towards the global coverage of a unified registry of IVF outcomes. Editorial. *Reproductive Biomedicine Online* 38(2): 133–137. Available at www.rbmojournal.com/article/S1472-6483(18)30598-4/pdf (Accessed 2 September 2023).

Felt, U., Fouché, R., Miller, C. et al. (eds.) (2017) *Handbook of Science and Technology Studies*. Cambridge, MA: MIT Press.

Fishkin, James (2018) *Democracy When the People Are Thinking*. Oxford: Oxford University Press.

  (2020) Cristina Lafont's challenge to deliberative minipublics. *Journal of Deliberative Democracy* 16(2): 56–62.

Fishkin, James., Mayega, Roy, Atuyambe, Lynn, Tumuhamye, Nathan, Ssentongo, Julius, Bazeyo, William, and Siu, Alice (2017) Applying deliberative democracy in Africa: Uganda's first deliberative polls. *Daedalus* 146(3): 140–154.

Floridi, L. (2014) *The Fourth Revolution – How the Infosphere Is Reshaping Human Reality*. Oxford: Oxford University Press.

  (2016) Faultless responsibility: On the nature and allocation of moral responsibility for distributed moral actions. *Royal Society's Philosophical Transactions A: Mathematical, Physical and Engineering Sciences* 374(2083): 1–22.

  (2017) Digital's cleaving power and its consequences. *Philosophy and Technology* 30(2): 123–129.

  (2020) Artificial intelligence as a public service: Learning from Amsterdam and Helsinki. *Philosophy & Technology* 33: 541–546.

Floridi, L., Taddeo M. and Turilli M. (2009) Turing's imitation game: Still an impossible challenge for all machines and some judges – An evaluation of the 2008 Loebner contest. *Minds and Machines* 19(1): 145–150.

Foa, Roberto and Mounk, Yascha (2019) Youth and the populist wave. *Philosophy and Social Criticism* 45(9–10): 1013–1024.

Forbes (2023) UK Artificial Intelligence AU Statistics and Trends in 2023. *Forbes* (June 2023). Available at UK Artificial Intelligence (AI) Statistics And Trends In 2023 (Accessed 8 October 2023).

Fotheringham, Ellen and Boswell, Caitlin (2022) 'Unequal Impacts': How UK immigration law and policy affected migrants' experiences of the Covid-19 pandemic. Public Interest Law Centre (May). Available at www.jcwi.org.uk/Handlers/Download.ashx?IDMF=6b9ce180-ef73-4d33-970f-88fa300feffd (Accessed 2 September 2023).

Franklin, Sarah (2013) The HFEA in context. *Reproductive Biomedicine Online* 26 (4): 310–312.

(2019a) Developmental landmarks and the Warnock Report: A sociological account of biological translation. *Comparative Studies in Society and History* 61 (4): 743–773.

(2019b) Obituary – Mary Warnock (1924–2019). *Nature* (17 April). Available at www.nature.com/articles/d41586-019-01277-5 (Accessed 2 September 2023).

(2019c) Ethical research – the long and bumpy road from shirked to shared. *Nature* 574: 627–630. Available at www.nature.com/articles/d41586-019-03270-4 (Accessed 2 September 2023).

Fraser, Nancy (2015) A feminism where 'lean in' means leaning on others. Interview with Gary Gutting, *The New York Times,* Opinionator (15 October). Available at https://archive.nytimes.com/opinionator.blogs.nytimes.com/2015/10/15/a-feminism-where-leaning-in-means-leaning-on-others/?smid=pl-share&_r=0&mtrref=www.google.com&gwh=C9D3F1DB07E24518D7869F6D92EB8DA4&gwt=regi&assetType=REGIWALL (Accessed 2 September 2023).

(2021) Gender, capital and care. In Jude Browne (ed.), *Why Gender?* Cambridge: Cambridge University Press, pp. 144–169.

Frey, Carl and Osborne, Michael (2013) The future of employment: How susceptible are jobs to computerization? Working Paper, Oxford: Oxford Martin.

Galston, William A. (2007) An old debate renewed: The politics of the public interest. *Daedalus* 136(4) Fall: 10–19.

Gash, T. and Rutter, J. (2011) The Quango Conundrum. *The Political Quarterly* 82: 95–101.

Gash, Tom, Magee, Ian, Rutter, Jill and Smith, Nicole (2010) *READ BEFORE BURNING: Arm's length government for a new administration.* London: Institute for Government. Available at www.instituteforgovernment.org.uk/sites/default/files/publications/Read%20before%20burning.pdf (Accessed 2 September 2023).

Gatrell, C., Ladge, J. J. and Powell, G. N. (2022). Profane pregnant bodies versus sacred organizational systems: Exploring pregnancy discrimination at work (R2). *Journal of Business Ethics.* https://doi.org/10.1007/s10551-023-05518-6.

Gavaghan, Colin (2007) *Defending the Genetic Supermarket: Law and Ethics of Selecting the Next Generation.* Abingdon: Routledge-Cavendish.

Gayer, Ted, Litan, Robert and Wallach, Philip (2017) *Evaluating the Trump Administration's Regulatory Reform Program.* Washington, DC: The Center on Regulation and Markets at the Brookings Institution. Available at www.brookings.edu/wp-content/uploads/2017/10/evaluatingtrumpregreform_gayerlitanwallach_102017.pdf (Accessed 2 September 2023).

Gill, Matthew and Dalton, Grant (2023) When should public bodies exist? Rewriting the 'three tests' for when government does things at arm's length. Institute for Government (July). Available at www.instituteforgovernment.org.uk/sites/default/files/2023–07/when-should-public-bodies-exist.pdf (Accessed 2 September 2023).

Global Labor Justice (2018) Gender-based violence in garment supply chains. Available at https://globallaborjustice.org/gap/ (Accessed 2 September 2023).

Goldman Sachs (2023) Generative AI could raise global GDP by 7%. Available at www.goldmansachs.com/intelligence/pages/generative-ai-could-raise-global-gdp-by-7-percent.html (Accessed 2 September 2023).

Goodhart, Michael (2017) Interpreting responsibility politically. *Journal of Political Philosophy* 25(2): 173–195.

(2018) *Injustice: Political Theory for the Real World.* Oxford: Oxford University Press.

Goodin, Robert E. (1998) Collective responsibility. In David Schmidtz and Robert E. Goodin (eds.), *Social Welfare and Individual Responsibility.* Cambridge: Cambridge University Press, pp. 145–168.

Goold, Imogen and Savulescu, Julian (2009) In favour of freezing eggs for non-medical reasons. *Bioethics* 23: 47–58.

Golombok, Susan. (2015) *Modern Families: Parents and Children in New Family Forms.* Cambridge: Cambridge University Press.

Gordon, Lewis (2007) Iris Marion Young on political responsibility: A reading through Jaspers and Fanon. *Symposia on Gender, Race and Philosophy* 3(1): 1–7. Available at http://web.mit.edu/sgrp/2007/no1/Gordon0107.pdf (Accessed 2 September 2023).

Greely, Henry (2017) Designer babies: an ethical horror waiting to happen? Interview by Philip Ball with Henry Greely, bioethicist of Stanford University in California. *The Guardian* (8 January). Available at www.theguardian.com/science/2017/jan/08/designer-babies-ethical-horror-waiting-to-happen (Accessed 2 September 2023).

Grzymala-Busse, Anna, Fukuyama, Francis, Kuo, Didi and McFaul, Michael (2020) Global Populisms and their Challenges. Stanford University White Paper: Stanford University. Available at https://stanford.app.box.com/s/0afiu4963qjy4gicahz2ji5x27tednaf (Accessed 2 September 2023).

Gürtin, Zeynep and Tiemann, Emily (2021) The marketing of elective egg freezing: A content, cost and quality analysis of UK fertility clinic websites. *Reproductive Biomedicine & Society Online* 12(4): 56–68.

Gyngell, Christopher, Bowman-Smart, Hilary and Savulescu, Julian (2019). Moral reasons to edit the human genome: picking up from the Nuffield Report. *Journal of Medical Ethics* 45(8): 514–523.

Habermas, Jürgen (1989 [1962]) *The Structural Transformation of the Public Sphere*, trans. Thomas Burger. Cambridge, MA: MIT Press.

(1990) Discourse ethics: Notes on a program of philosophical justification. In Seyla Benhabib and Fred Dallmayer (eds.), *The Communicative Ethics Controversy*. Cambridge MA: MIT Press, pp. 60–110.

(2003) *The Future of Human Nature*. Oxford: Polity Press.

Hagendroff, Thilo (2020) The ethics of AI ethics: An evaluation of guidelines. *Minds and Machines* 30: 99–120.

Halberstam, Jack (2021) Gender and the Queer/Trans* Undercommons. In Jude Browne (ed.), *Why Gender?* Cambridge: Cambridge University Press, pp. 38–56.

Haldane, Andrew (2015) Speech to the Trades Union Congress, London (12 November). Available at www.bankofengland.co.uk/-/media/boe/files/speech/2015/labours-share.pdf?la=en&hash=D6F1A4C489DA855C8512FC41C02E014F8D683953 (Accessed 2 September 2023).

Hanfeld, Michael (2023) More than 200 positions will be lost for 'Bild'. *FrankFurter Allgemeine* (19 June). Available at www.faz.net/aktuell/feuilleton/medien/bild-zeitung-entlaesst-mehr-als-200-mitarbeiter-ki-haelt-einzug-18974856.html (Accessed 2 September 2023).

Harari, Yuval Noah (2017) *Homo Deus: A Brief History of Tomorrow*. London: Vintage Books.

(2021). Will artificial intelligence create useless class of people? *Channel 4 News* (29 October). Online video clip. Available at www.youtube.com/watch?v=7FzNUc-ZFv4 (Accessed 2 September 2023).

Haraway, D. (2000 [1985]) A cyborg manifesto: Science, technology and socialist-feminism in the late twentieth century. In G. Kirkup, L. Janes, F. Hovenden et al. (eds.), *The Gendered Cyborg: A Reader*. London: Routledge, pp. 50–57.

Haraway, Donna (1988) Situated knowledges: The science question in feminism and the privilege of partial perspective. *Feminist Studies* 14(3): 575–599.

Harding, Verity (forthcoming 2024) *AI Needs You.* Princeton, NJ: Princeton University Press.

Harries, Lord Harries of Pentregarth (2010) Public Bodies Bill (9 November 2010). House of Lords, Hansard Text: Column 111. Available from: www.publications.parliament.uk/pa/ld201011/ldhansrd/text/101109–0001.htm#10110950000414.

Harris, John (2009) Enhancements are a moral obligation. In Julian Savulescu and Nick Bostrom (eds.), *Human Enhancement.* Oxford: Oxford University Press, pp. 131–154.

　(2017) Should we genetically modify humans? Interview with Professor John Harris. Practical Ethics Channel. Online video clip. Available at www .youtube.com/watch?v=kIMOpu9uZbk (Accessed 2 September 2023).

Harris, Tristan (2020) Interview in the Netflix Documentary – The Social Dilemma. Available at www.thesocialdilemma.com/ (Accessed 2 September 2023).

Harvard Law Review (2017) Criminal Law – Sentencing guidelines – Wisconsin Supreme Court requires warning before use of algorithmic risk assessments in sentencing. – State v. Loomis, 881 N.W.2d 749 (Wis. 2016). *Recent Cases* 130: 1530–1537. Available at https://harvardlawreview.org/wp-content/uploads/2017/03/1530-1537_online.pdf (Accessed 2 September 2023).

Haslanger, Sally (2015) What is a (social) structural explanation? *Canadian Journal of Philosophy* 45(1): 1–15.

　(2016) What is a (social) structural explanation? *Philosophical Studies* 173 (1): 113–130.

　(2022) How to change a social structure. Available at www.ucl.ac.uk/laws/sites/laws/files/haslanger_how_to_change_a_social_structure_ucl.pdf (Accessed 2 September 2023).

　(2024) Agency under structural constraints in social systems. In Jude Browne and Maeve McKeown (eds.), *What Is Structural Injustice?* Oxford: Oxford University Press, pp. 48–64.

Hassan, Katie (2020) Gene editing is not safe, not needed & unethical. The Future of Genetics. Doha Debates (28 May). Available at https://dohadebates.com/video/julian-savulescu-genetic-enhancement-is-a-moral-obligation/ (Accessed 7 September 2023).

Hayles, K. (1999) *How We Became Posthuman: Virtual Bodies in Cybernetics, Literature, and Informatics.* Chicago, IL: University of Chicago Press.

Hayles, Katherine (2023) Technosymbiosis: Figuring (out) our relations to AI. In Jude Browne et al. (eds.), *Feminist AI: Critical Perspectives on Algorithms, Data, and Intelligent Machines.* Oxford: Oxford University Press, pp. 1–18.

Hayward, Clarissa (2017) Responsibility and ignorance: On dismantling structural injustice. *The Journal of Politics* 79(2): 396–408.

Heidegger, Martin (1977) The question concerning technology. In *The Question Concerning Technology and Other Essays*. New York: Garland Publishing, pp. 3–35.

Helberger, Natali, and Diakopoulos, Nicholas (2023). ChatGPT and the AI Act. *Internet Policy Review* 12(1). Available at https://policyreview.info/essay/ chatgpt-and-ai-act (Accessed 2 September 2023).

HFEA (2013) *Mitochondria Replacement Consultation: Advice to Government*. London: HFEA.

(2015) World first as mitochondrial donation regulations come into force. Available at https://bit.ly/3OGLEmB (Accessed 2 September 2023).

(2017) *Innovation in Regulation*. London: HFEA. Available at http://ifqtesting .blob.core.windows.net/umbraco-website/1351/hfea-innovation-and-regula tion-plan-feb-2017.pdf (Accessed 9 September 2023).

(2020) Storage consultation response. Available at www.hfea.gov.uk/media/ 3102/22-04-2020-storage-consultation-response.pdf (Accessed 2 September 2023).

(2023a) Fertility treatment 2021: Preliminary trends and figures (June). Available at www.hfea.gov.uk/about-us/publications/research-and-data/fer tility-treatment-2021-preliminary-trends-and-figures/#main-points (Accessed 2 September 2023).

(2023b) Approved PGT-M and PTT conditions. Available at www.hfea.gov.uk/ treatments/embryo-testing-and-treatments-for-disease/approved-pgt-m-and-ptt-conditions/ (Accessed 2 September 2023).

(2023c) Buisness Plan 2022–2023. Available at www.hfea.gov.uk/media/ ve1ce035/hfea-business-plan-2022–2023.pdf (Accessed 9 September 2023).

Hill Collins, Patricia (2017) The difference that power makes: Intersectionality and participatory democracy. *Investigaciones Feministas* 8(1): 19–39.

(2019) *Intersectionality as Critical Social Theory*. Durham, NC: Duke University Press, 2019.

Hill Collins, Patricia and Bilge, Sirma (2020) *Intersectionality: Key Concepts*. Cambridge: Polity Press.

Hillel, David, McCague, Paul W. and Yaniszewski Peter F. (2005) Proving caus-ation where the but for test is unworkable. *The Advocates Quarterly* 30: 216–238.

HM Government (2015) Annex: Public Bodies Reform Programme Update, 1 December 2015. Available at www.gov.uk/government/uploads/system/ uploads/attachment_data/file/506070/Annex_-_Public_Bodies_Reform_ Programme_Update_16_Dec_2015.pdf (Accessed 2 September 2023).

(2018) Government response to House of Lords Artificial Intelligence Select Committee's Report on AI in the UK: Ready, Willing and Able? (June).

Available at www.parliament.uk/globalassets/documents/lords-commit
tees/Artificial-Intelligence/AI-Government-Response.pdf (Accessed 2
September 2023).

(2020a) Revolutionary Artificial Intelligence warship contracts announced (14
January). Available at www.gov.uk/government/news/revolutionary-artificial-
intelligence-warship-contracts-announced (Accessed 2 September 2023).

(2020b) Artificial Intelligence and Public Standards: Report (10 February).
Available at www.gov.uk/government/publications/artificial-intelligence-
and-public-standards-report (Accessed 2 September 2023).

(2021a) National AI Strategy. Available at www.gov.uk/government/publica
tions/national-ai-strategy (Accessed 2 September 2023).

(2021b) AI Roadmap (6 January). Available at www.gov.uk/government/publi
cations/ai-roadmap (Accessed 2 September 2023).

(2021c) National AI Strategy (22 September). Available at www.gov.uk/govern
ment/publications/national-ai-strategy (Accessed 2 September 2023).

(2021d) Free courses for jobs (1 April). Available at www.gov.uk/guidance/
free-courses-for-jobs (Accessed 2 September 2023).

(2022a) Establishing a pro-innovation approach to regulating AI (updated 20 July).
Available at www.gov.uk/government/publications/establishing-a-pro-innov
ation-approach-to-regulating-ai/establishing-a-pro-innovation-approach-to-regu
lating-ai-policy-statement (Accessed 2 September 2023).

(2022b) Press statement: UK sets out proposals for new AI rulebook to unleash
innovation and boost public trust in the technology (18 July). Available at www
.gov.uk/government/news/uk-sets-out-proposals-for-new-ai-rulebook-to-unleash-
innovation-and-boost-public-trust-in-the-technology (Accessed 2 September 2023).

(2022c) Launching the AI Research Centre for Defence (14 July). Available at
www.gov.uk/government/news/launching-the-defence-centre-for-ai-research
(Accessed 2 September 2023).

(2023a) A pro-innovation approach to AI regulation (3 August). Available at www
.gov.uk/government/publications/ai-regulation-a-pro-innovation-approach/
white-paper (Accessed 2 September 2023).

(2022d) National AI Strategy – AI Action Plan (18 July). Available at www.gov
.uk/government/organisations/office-for-artificial-intelligence (Accessed
2 September 2023).

(2023b) UK Science and Technology Framework Policy Paper. Available at www
.gov.uk/government/publications/uk-science-and-technology-framework
(Accessed 2 September 2023).

(2023c) Chancellor reveals life sciences growth package to fire up economy.
Available at www.gov.uk/government/news/chancellor-reveals-life-sciences-
growth-package-to-fire-up-economy (Accessed 2 September 2023).

(2023d) Seven principles of public office. Available at www.gov.uk/govern ment/publications/the-7-principles-of-public-life/the-7-principles-of-public-life—2 (Accessed 9 September 2023).

Horsey, Kirsty (ed.) (2015) *Revisiting the Regulation of Human Fertilisation and Embryology.* London: Routledge.

House of Commons Library (2022) Ethnic diversity in politics and public life. Available at https://commonslibrary.parliament.uk/research-briefings/ sn01156/ (Accessed 2 September 2023).

(2023a) Women in politics and public life. Available at https:// commonslibrary.parliament.uk/research-briefings/sn01250/ (Accessed 2 September 2023).

(2023b) Adult Social Care Funding (England). Available at https:// commonslibrary.parliament.uk/research-briefings/cbp-7903/ (Accessed 2 September 2023).

(2020) Killer robots: Should lethal autonomous weapons be banned? (20 November). Available at https://lordslibrary.parliament.uk/killer-robots-should-lethal-autonomous-weapons-be-banned/ (Accessed 2 September 2023).

House of Lords Select Committee (2018) AI in the UK: Ready, willing and able? Available at https://publications.parliament.uk/pa/ld201719/ldselect/ ldai/100/10013.htm (Accessed 2 September 2023).

Human Fertilisation and Embryology Authority (HFEA) (13 November 2013) *Authority Agenda Wednesday.* London: HFEA.

Huntingdon, Samuel, P. (1953) The Marasmus of ICC: The Commission, the Railroads and the Public Interest. *Yale Law Journal* 61: 467–509.

Inhorn, Marcia (2023) *Motherhood on Ice: The Mating Gap and why Women Freeze their Eggs.* New York: New York University Press.

Inhorn, Marcia, Carmeli, Birenbaum-Daphna, Birger J. et al. (2018) Elective egg freezing and its underlying socio-demography: A binational analysis with global implications. *Reproductive Biology and Endocrinology* 16(1): 1–11.

IfG – Institute for Government (2010) *Read Before Burning: Arms' Length Government for a New Administration.* London: IfG.

Involve (2018) Citizens' Assembly on social care; Recommendations for funding adult social care. Report. Available at www.involve.org.uk/sites/default/files/ field/attachemnt/Citizens%27%20Assembly%20on%20Social%20Care%20-%20Recommendations%20for%20funding%20social%20care_0.pdf (Accessed 2 September 2023).

Janeway, Bill (2018) American political economy, disrupted. Available at www .billjaneway.com/american-political-economy-disrupted (Accessed 2 September 2023).

JDSUPRA (2021) European Commission's Proposed Regulation on Artificial Intelligence: Requirements for High-Risk AI Systems (18 June). Available at www.jdsupra.com/legalnews/european-commission-s-proposed-3823933/ (Accessed 2 September 2023).

Jobin, Anna, Ienca, Marcello and Vayena, Effy (2019) The global landscape of AI Ethics guidelines. *Nature Machine Intelligence* 1(9): 389–399.

Johnson, Boris (2019) PM speech to the UN General Assembly (24 September). Available at www.gov.uk/government/speeches/pm-speech-to-the-un-gen eral-assembly-24-september-2019 (Accessed 2 September 2023).

Johnson, James (2020) Boris Johnson risks dragging Brits out of lockdown against their will. Politico (12 May). Available at www.politico.eu/article/boris-johnson-risks-dragging-brits-out-of-coronavirus-lockdown-against-their-will/ (Accessed 2 September 2023).

Johnson, Simon, and Kwak, James (2010) *13 Bankers: The Wall Street Takeover and the New Financial Meltdow.* New York: Panthleon, pp 104–105.

Jones, Phil (2021) *Work without the Worker: Labour in the Age of Platform Capitalism.* London: Verso.

Jones, P. (2022) Artificial intelligence quietly relies on workers earning $2 per hour. *Sciencefocus* (7 January). Available at www.sciencefocus.com/future-tech nology/artificial-intelligence-quietly-relies-on-workers-earning-2-per-hour/ (Accessed 2 September 2023).

Judd, Lord Judd (9 November 2010) Public Bodies Bill. House of Lords, Hansard Text: Column 141. Available at www.publications.parliament.uk/pa/ ld201011/ldhansrd/text/101109–0001.htm#10110950000414.

Jugov, Tamara and Ypi, Lea (2019) Structural injustice, epistemic opacity, and the responsibilities of the oppressed. *Journal of Social Philosophy* 50(1): 7–27.

Kant, Immanuel (1959 [1797]) *Foundations of the Metaphysics of Morals,* trans. Lewis White Beck, Library of Liberal Arts. Indianapolis, IN: Bobbs-Merrill.

Kasirzadeh, Atoosa (2022) Algorithmic fairness and structural injustice: Insights from feminist political philosophy. Conference paper, Fifth AAAI/ACM Conference on Artificial Intelligence, Ethics and Society (AIES), 1–3 August, Oxford. Available at https://bit.ly/3sdqGEf (Accessed 2 September 2023).

Katz, Cindi (2021) Aspiration management: Gender, race, class, and the child as waste. In Jude Browne (ed.), *Why Gender?* Cambridge: Cambridge University Press, pp. 170–193.

Kayser-Bril, Nicolas (2020) Google apologizes after its Vision AI produced racist results. Algorithm Watch (8 April). Available at https://algorithmwatch .org/en/google-vision-racism/ (Accessed 2 September 2023).

Keynes, John Maynard (1951 [1930]) Economic possibilities for our grandchildren. In *Essays in Persuasion.* London: Rupert Hart-Davis, pp. 358–373.

King, Sir Mervyn, Governor of the Bank of England (2011). Addressing the Draft Financial Services Bill (18 October). Available at www.telegraph.co.uk/finance/newsbysector/banksandfinance/8868041/Bank-of-England-should-be-left-alone-to-police-banks-says-Governor-Sir-Mervyn-King.html.

Knight, W. (2017) The dark secret at the heart of AI. MIT Technology Review. Available at www.technologyreview.com/2017/04/11/5113/the-dark-secret-at-the-heart-of-ai/ (Accessed 2 September 2023).

Kozyrkov, Cassie (2020) What is 'ground truth' in AI? (A warning.) Towards Data Science (28 February). Available at https://towardsdatascience.com/in-ai-the-objective-is-subjective-4614795d179b (Accessed 2 September 2023).

Kuhnt, A. K. and Passet-Wittig, J. (2022) Families formed through assisted reproductive technology: Causes, experiences, and consequences in an international context. *Reproductive Biomedicine & Society Online* 14: 289–296.

Kutz, Christopher (2000) *Complicity: Ethics and Law for a Collective Age.* Cambridge: Cambridge University Press.

Kwak, James (2014) Cultural capture and the financial crisis. In David Carpenter and David, A. Moss (eds.) *Preventing Regulatory Capture: Special Interest Influence and How to Limit It.* Cambridge: Cambridge University Press, pp 71–98.

Laidlaw, James (2003) A free gift makes no friends. *Journal of the Royal Anthropological Institute.* 6(4): 617–634.

Lafont, C. (2015) Deliberation, participation, and democratic legitimacy: Should deliberative minipublics shape public policy? *Journal of Political Philosophy* 23(1): 40–63.

(2019) *Democracy without Shortcuts: A Participatory Conception of Deliberative Democracy.* Oxford: Oxford University Press.

Lavanchy, M. (2018) Amazon's sexist hiring algorithm could still be better than a human. *Quartz* (7 November). Available at https://qz.com/work/1454396/amazons-sexist-hiring-algorithm-could-still-be-better-than-a-human/ (Accessed 2 September 2023).

Lawrence, Neil (2015) The information barons threaten our autonomy and our privacy. *The Guardian* (16 November). Available at www.theguardian.com/media-network/2015/nov/16/information-barons-threaten-autonomy-privacy-online (Accessed 2 September 2023).

Levene, Mark (2022) Uncertainty quantification in artificial intelligence and machine learning. The Alan Turing Institute. Available at https://aistandardshub.org/uncertainty-quantification-artificial-intelligence-machine-learning (Accessed 2 September 2023).

Lever, Annabelle (2013) Democracy deliberation and public policy reform. In Henry Kippin and Gerry Stocker (eds.), *Public Services: A New Reform Agenda.* London: Bloomsbury Academic Press, pp. 91–106.

Levi-Faur, David (ed.) (2011) Regulation and regulatory governance. In *Handbook on the Politics of Regulation*. Northampton, MA: Edward Elgar Publishing, pp. 3–24.

Loane, Maria, Morris, Joan K., Addor, Marie-Claude, Arriola, Larraitz, Budd, Judith, Doray, Berenice, Ester, Garne et al. (2013). Twenty-year trends in the prevalence of Down syndrome and other trisomies in Europe: Impact of maternal age and prenatal screening. *European Journal of Human Genetics* 21 (1): 27–33.

Lu, Catherine (2017) *Justice and Reconciliation in World Politics*. Cambridge: Cambridge University Press.

(2018) Responsibility, structural injustice, and structural transformation. *Ethics and Global Politics* 11(1): 42–57.

Maboloc, Christopher (2019) What is structural injustice? *Philosophia* 47: 1185–1196.

MacAskill, William (2022) *What We Owe the Future: A Million-Year View*. London: Oneworld Publications.

Macron, Emmanuel (2018) Emmanuel Macron talks to WIRED about France's AI strategy. Interview with Nicholas Thompson. *Wired* (31 March). Available at www.wired.com/story/emmanuel-macron-talks-to-wired-about-frances-ai-strategy/ (Accessed 2 September 2023).

Mansbridge, Jane and Boot, Eric (2022). Common Good. In International Encyclopedia of Ethics. https://doi.org/10.1002/9781444367072.wbiee608.pub2.

Manthrope, Roland (2017) To make a new kind of shoe, Adidas had to change everything. *Wired* (4 October). Available at www.wired.co.uk/article/adidas-speedfactory-made-for-london-trainers (Accessed 2 September 2023).

Marin, Mara (2018) Racial structural solidarity. *Critical Review of International Social and Political Philosophy* 21(5): 586–600.

Marx, Karl (1990 [1867]) *Capital: Critique of Political Economy*, vol. 1. London: Penguin Classics.

Maude, Francis (2010) Public Body Review Published (14 October). London: HMG. Available at www.gov.uk/government/news/public-body-review-pub lished (Accessed 2 September 2023).

(2011a) Quango reforms take a leap forward as Public Bodies Act receives Royal Assent. Cabinet Office Press Release (15 December). London: HMG. Available at www.gov.uk/government/news/quango-reforms-take-a-leap-for ward-as-public-bodies-act-receives-royal-assent (Accessed 2 September 2023).

(2011b) Written Ministerial Statement Public Bodies Act 2011. Cabinet Office (15 December). London: HMG. Available at www.gov.uk/government/ uploads/system/uploads/attachment_data/file/62124/Written_Ministerial_ Statement_Public_Bodies_Act_2011.pdf (Accessed 2 September 2023).

Mauss, Marcel (1990 [1954]). *The Gift: The Form and Reason for Exchange in Archaic Societies.* Routledge: London.

McCarty, Nolan (2014) Complexity, capacity and capture. In David Capenter and David, A. Moss (eds.), *Preventing Regulatory Capture: Special Interest Influence and How to Limit It.* Cambridge: Cambridge University Press, pp 99–123.

McKee, Rebecca (2018) The Irish Abortion Referendum. Blog (30 May). Available at www.involve.org.uk/resources/blog/opinion/citizens-assembly-behind-irish-abortion-referendum (Accessed 2 September 2023).

McKeown, Maeve (2018) Iris Marion Young's 'social connection model' of responsibility: Clarifying the meaning of connection. *Journal of Social Philosophy* 49(3): 484–502.

(2024) Pure, avoidable, and deliberate structural injustice. In Jude Browne and Maeve McKeown (eds.), *What is Structural Injustice?* Oxford: Oxford University Press, pp. 65–84.

McNamee, Roger (2020) Interview in the Netflix Documentary – The Social Dilemma. Available at www.thesocialdilemma.com/ (Accessed 2 September 2023).

Metzl, Jamie (2019) *Hacking Darwin: Genetic Engineering and the Future of Humanity.* Naperville, IL: Sourcebooks.

Midland Fertility Clinic (2012). The MFS 25th Anniversary 25 Milestone Babies. Tamworth. MFS. Available at www.midlandfertility.com/2012/07/newsletter-issue-25-made-in-aldridge/ (Accessed 2 September 2023).

Microsoft (2019) 2019 Manufacturing Trends Report. Available at http://info.microsoft.com/rs/157-GQE-382/images/EN-US-CNTNT-Report-2019-Manufacturing-Trends.pdf (Accessed 2 September 2023).

Miller, David (2007) *National Responsibility and Global Justice.* Oxford: Oxford University Press.

(2016) *Strangers in Our Midst.* Cambridge, MA: Harvard University Press.

Milmo, Dan and Hern, Alex (2023) Discrimination is a bigger AI risk than human extinction – EU commissioner. *The Guardian* (14 June). Available at www.theguardian.com/technology/2023/jun/14/ai-discrimination-is-a-bigger-risk-than-human-extinction-eu-chief (Accessed 2 September 2023).

Minds at Work (2018) Artificial Intelligence struggles to distinguish a blueberry muffin from a chihuahua, so that's something we still do better than robots (12 December). Available at www.mindsatwork.com.au/tweet_post/artificial-intelligence-struggles-to-distinguish-a-blueberry-muffin-from-a-chihuahua-so-thats-something-we-still-do-better-than-robots/ (Accessed 2 September 2023).

Ministry of Defence (2017) Unmanned Aircraft Systems (September). Available at www.gov.uk/government/publications/unmanned-aircraft-systems-jdp-0-302 (Accessed 2 September 2023).

Ministry of Defence (2022) AMBITIOUS, SAFE, RESPONSIBLE Our approach to the delivery of AI enabled capability in Defence (June). Available at https://assets.publishing.service.gov.uk/government/uploads/system/uploads/attachment_data/file/1082991/20220614-Ambitious_Safe_and_Responsible.pdf (Accessed 2 September 2023).

Minteer, Ben (2005) Environmental philosophy and the public interest: A pragmatic reconciliation. *Environmental Values* 14(1): 37–60.

Mitchell, Melanie (2019) We Shouldn't be Scared by 'Superintelligent A.I.' 'Superintelligence' is a flawed concept and should not inform our policy decisions. *The New York Times* (31 October). www.nytimes.com/2019/10/31/opinion/superintelligent-artificial-intelligence.html (Accessed 2 September 2023).

Mitchell, Tom and Brynjolfsson, Eric (2017) Track how technology is transforming work. *Nature* 544: 290–292. Available at www.cs.cmu.edu/~tom/pubs/Nature2017_Mitchell_Brynjolfsson_FINAL.pdf (Accessed 2 September 2023).

Moffitt, Benjamin (2020) *Populism: Key Concepts in Political Theory*. Cambridge: Polity Press.

Money-Coutts, Sophie (2021). Could freezing eggs be the new 21st birthday gift? *The Times* (7 September). Available at www.thetimes.co.uk/article/could-freezing-eggs-be-the-new-21st-birthday-gift-6x6l2l8jr (Accessed 2 September 2023).

Mounk, Yascha (2019) *The People vs. Democracy*. Cambridge, MA: Harvard University Press.

Muller, Michael, Wolf, Christine, Andres, Josh et al. (2021) Designing ground truth and the social life of labels. ACM Reference Format. Available at www.researchgate.net/publication/348416620_Designing_Ground_Truth_and_the_Social_Life_of_Labels_ACM_Reference_Format (Accessed 2 September 2023).

Musk, Elon (2023) quoted by Dan Milmo (2023) Elon Musk launches AI startup and warns of a 'Terminator future'. *Guardian Newspaper* (13 July). Available at www.theguardian.com/technology/2023/jul/13/elon-musk-launches-xai-startup-pro-humanity-terminator-future (Accessed 9 October 2023).

Nagel, Thomas (1974) What is it like to be a bat? *The Philosophical Review* 83(4): 435–450.

Needham, Carol A. (2010) Listening to Cassandra: 'The Difficulty of Recognizing Risks and Taking Action'. *Fordham Law Review* 78: 2347–2355.

Nielly, C. (2020) Can we let algorithm take decisions we cannot explain? *Towards Data Science* (12 February). Available at https://towardsdatascience.com/can-we-let-algorithm-take-decisions-we-cannot-explain-a4e8e51e2060 (Accessed 2 September 2023).

Niemeyer, Simon and Jennstål, Julia (2018) Scaling up deliberative effects – applying lessons of mini-publics. In André Bächtiger, John S. Dryzek, Jane J. Mansbridge and Mark Warren (eds.), *The Oxford Handbook of Deliberative Democracy*. Oxford: Oxford University Press, pp. 329–347.

Noble, S. U. (2018) *Algorithms of Oppression: How Search Engines Reinforce Racism*. New York: New York University Press.

Nozick, Robert (1974) *Anarchy, State, and Utopia*. New York: Basic Books.

Nussbaum, Martha (2009) Iris Young's last thoughts on responsibility for global justice. In Ann Ferguson and Nagel Mechthild (eds.), *Dancing with Iris: The Philosophy of Iris Marion Young*. Oxford: Oxford University Press, pp. 133–146.

   (2011) Foreword. In Iris Marion Young (ed.), *Responsibility for Justice*. Oxford: Oxford University Press, pp. iv–xxv.

Nuti, Alasia (2019) *Injustice and the Reproduction of History: Structural Inequalities, Gender and Redress*. Cambridge: Cambridge University Press.

Obama, Michelle (2020) quoted in Rico, Klaritza (2020) Michelle Obama's graduation speech encourages activism beyond hashtags and posts. Variety website (7 June). Available at https://variety.com/2020/digital/news/michelle-obama-graduation-speech-dear-class-of-2020-1234626792-1234626792/ (Accessed 2 September 2023).

O'Carroll, Lisa (2023) EU moves closer to passing one of world's first laws governing AI. *The Guardian* (14 June). Available at www.theguardian.com/technology/2023/jun/14/eu-moves-closer-to-passing-one-of-worlds-first-laws-governing-ai (Accessed 2 September 2023).

O'Cinneide, Colm (forthcoming 2024) Cruel optimism? Structural justice and the blunted promise of equality law. In Virginia Mantouvalou and Jonathan Wolf (eds.), *Structural Injustice and the Law*. London: UCL Press.

O'Neil, Cathy (2016) *Weapons of Math Destruction*. New York: Crown Publishing Group.

O'Neill, Onora (1988) Hunger, needs and rights. In Steven Luper-Foy (ed.), *Problems of International Justice*. London: Routledge, pp. 67–83.

   (2000) *Bounds of Justice*. Cambridge: Cambridge University Press.

   (2002) *Autonomy and Trust in Bioethics*. Cambridge: Cambridge University Press.

   (2005) The dark side of human rights. *International Affairs* 81(2): 427–439.

OECD (2019) Recommendation of the Council on Artificial Intelligence (5 May). Available at https://legalinstruments.oecd.org/en/instruments/OECD-LEGAL-0449 (Accessed 2 September 2023).

   (2020) AI Strategies & Public Sector Components. Available at https://oecd-opsi.org/projects/ai/strategies/ (Accessed 2 September 2023).

   (2021) National AI Policies & Strategies. Available at https://oecd.ai/dashboards?selectedTab=countries (Accessed 2 September 2023).

Office for Artificial Intelligence (2023) White Paper – A Pro-innovation Approach to AI Regulation. Available at www.gov.uk/government/publications/ai-regulation-a-pro-innovation-approach/white-paper (Accessed 9 April 2024).

Office for National Statistics (ONS) (2018) Human capital (1 October). Available at www.ons.gov.uk/releases/humancapital (Accessed 2 September 2023).

(2019a) Average weekly earnings in Great Britain (July 2019). Available at www.ons.gov.uk/employmentandlabourmarket/peopleinwork/employmentandemployeetypes/bulletins/averageweeklyearningsingreatbritain/july2019 (Accessed 2 September 2023).

(2019b) Which occupations are at highest risk of being automated? (25 March). Available at www.ons.gov.uk/employmentandlabourmarket/peopleinwork/employmentandemployeetypes/articles/whichoccupationsareathighestriskofbeingautomated/2019-03-25 (Accessed 30 August 2022) (Accessed 2 September 2023).

(2020) Labour market economic commentary (March). Available at www.ons.gov.uk/employmentandlabourmarket/peopleinwork/employmentandemployeetypes/articles/labourmarketeconomiccommentary/latest (Accessed 2 September 2023).

(2021) EMP17: People in employment on zero hours contracts 2000–2020. Available at www.ons.gov.uk/employmentandlabourmarket/peopleinwork/employmentandemployeetypes/datasets/emp17peopleinemploymentonzerohourscontracts (Accessed 2 September 2023).

(2022b) Overview of human capital estimates in the UK: 2004 to 2020 (25 April). Available at www.ons.gov.uk/peoplepopulationandcommunity/wellbeing/articles/humancapitalestimates/2004to2020 (Accessed 2 September 2023).

(2023a) Employment in UK: August 2023. Available at www.ons.gov.uk/employmentandlabourmarket/peopleinwork/employmentandemployeetypes/bulletins/employmentintheuk/latest (Accessed 3 September 2023).

(2023b) Average weekly earnings in Great Britain: August 2023 Estimates of growth in earnings for employees before tax and other deductions from pay. Available at www.ons.gov.uk/employmentandlabourmarket/peopleinwork/employmentandemployeetypes/bulletins/averageweeklyearningsingreatbritain/april2023#:~:text=Average%20weekly%20earnings%20were%20estimated, (COVID%2D19)%20pandemic (Accessed 2 September 2023).

Parekh, Serena (2011) Getting to the root of gender inequality: Structural injustice and political responsibility. *Hypatia* 26(4): 672–689.

Parfit, Derek (1984) *Reasons and Persons.* Oxford: Clarendon Press.

Park S. U., Walsh, L. and Berkowitz, K. M. (2021) Mechanisms of ovarian aging. *Reproduction* 162(2): R19–R33.

Parsons, Alex (2019) Digital Tools for Citizens' Assemblies. My Society (June). Available at https://research.mysociety.org/media/outputs/digital-tools-citizens-assemblies.pdf (Accessed 2 September 2023).

Patel, Nilay (2018) Mark Zuckerberg is 'actually not sure we should not be regulated'; And an immediate pivot to self-regulation, of course'. *The Verge* (22 March). Available at www.theverge.com/2018/3/21/17150270/mark-zuckerberg-facebook-regulated (Accessed 2 September 2023).

Peebles, Gustav (2010). The anthropology of credit and debt. *Annual Review of Anthropology* 39: 225–240.

Peltzman, Sam (1976) Toward a more general theory of regulation. *Journal of Law and Economics* 19: 211–240.

Persson, Ingmar and Savulescu, Julian (2019) The duty to be morally enhanced. *Topoi* 38: 7–14.

Pettit, Philip (2007) Responsibility incorporated. *Ethics* 117: 171–201.

Peyton-Jones, T (2024) The Future of Work: Jobs and Skills in 2030. Available at https://assets.publishing.service.gov.uk/media/5a7dd8e1e5274a5eaea66b20/the_future_of_work_key_findings_edit.pdf

Phillips, Anne (2012) Representation and inclusion. *Politics & Gender* 8(4): 512–518.

(2015) *The Politics of the Human.* Cambridge: Cambridge University Press.

(2016). Interview with Eva Wiseman: We need to talk about egg freezing. *The Guardian* (7 February). Available from www.theguardian.com/society/2016/feb/07/life-on-hold-with-frozen-eggs (Accessed 3 September 2023).

(2020) Descriptive representation revisited. In Robert Rohrschneider and Jacques Thomassen (eds.), *The Oxford Handbook of Political Representation in Liberal Democracies.* Oxford: Oxford University Press, pp. 174–191.

(2021) *Unconditional Equals.* Princeton, NJ: Princeton University Press.

Pinkert, Felix (2014) What we together can (be required to) do. *Midwest Studies in Philosophy* XXXVIII: 187–202.

Pitkin, Hanna (1967) *The Concept of Representation.* Berkeley: University of California Press.

Posner, Richard (2014) A short, inglorious history. In David Capenter and David, A. Moss (eds.), *Preventing Regulatory Capture: Special Interest Influence and How to Limit It.* Cambridge: Cambridge University Press, pp. 49–56.

Poulton, Joanna (2016) Three-parent baby raises issues of long-term health risks. University of Oxford website. Available at www.ox.ac.uk/research/three-parent-baby-raises-issues-long-term-health-risks (Accessed 2 September 2023).

Powers, Madison and Faden, Ruth (2019) *Structural Injustice: Power, Advantage and Human Rights.* Oxford: Oxford University Press.

Price, Cedric (1966) quoted by the Royal Society of Arts. Available at www
.royalacademy.org.uk/event/technology-answer-what-question (Accessed
on 13 September 2023).

Putin, Vladimir (2007) quoted in Vincent, James (2017) Putin says the nation
that leads in AI 'will be the ruler of the world'. *The Verge* (4 September).
Available at www.theverge.com/2017/9/4/16251226/russia-ai-putin-rule-
the-world (Accessed 2 September 2023).

PwC (2017) Sizing the prize: What's the real value of AI for your business and
how can you capitalise? Available at www.pwc.com/gx/en/issues/analytics/
assets/pwc-ai-analysis-sizing-the-prize-report.pdf (Accessed 2 September
2023).

(2018) The macroeconomic impact of artificial intelligence. Available at www
.pwc.co.uk/economic-services/assets/macroeconomic-impact-of-ai-technical-
report-feb-18.pdf (Accessed 2 September 2023).

(2020) An introduction to implementing AI in manufacturing. Available at
www.pwc.com/gx/en/industrial-manufacturing/pdf/intro-implementing-ai-
manufacturing.pdf (Accessed 2 September 2023).

(2022) How will automation impact jobs? Available at www.pwc.co.uk/services/
economics/insights/the-impact-of-automation-on-jobs.html (Accessed 2
September 2023).

Rawls, John (1971) *A Theory of Justice*. Cambridge, MA: Harvard University Press.

Reiman, Jeffrey (2012) The structure of structural injustice: thoughts on Iris
Marion Young's Responsibility for Justice. *Social Theory and Practice* 38(4):
738–751.

Reingold, Beth (2019) Gender, race/ethnicity, and representation in state legis-
latures. *Political Science and Politics* 52(3): 426–429.

Rodriguez-Wallberg, Kenny and Kutluk, Oktay (2012). Recent advances in oocyte
and ovarian tissue cryopreservation and transplantation. *Best Practice Research
Clinical Obstetrics and Gynaecology* 26(3): 391–405.

Rorty, Richard (1993) Human rights, rationality, and sentimentality. In Stephen
Shute and Susan Hurley (eds.), *On Human Rights: The Oxford Amnesty Lectures*.
New York: Basic Books, pp. 111–134.

Rousseau, Jean-Jacques (1997 [1762]) *'The Social Contract' and Other Later Political
Writings*, Victor Gourevitch (ed.). Cambridge: Cambridge University Press.

Roy, T. Tsao (2004) Arendt and the modern state: Variations on Hegel in 'The
Origins of Totalitarianism'. *The Review of Politics* 66(1): 105–136.

Royal Society (2018) *The Impact of Artificial Intelligence on Work*. London: The Royal
Society. Available at https://royalsociety.org/topics-policy/projects/ai-and-
work/ (Accessed 2 September 2023).

Ruby-Merlin, P. and Jayam, R. (2018) Artificial intelligence in human resource management. *International Journal of Pure and Applied Mathematics* 119(17): 1891–1895.

Runciman, David (2007) The paradox of representation. *Journal of Political Philosophy* 15(1): 93–114.

(2018) *How Democracy Ends*. London: Profile Books.

(2019) Under the Skin Podcast (14 April). Available at www.youtube.com/watch?v=2dLN3cvOEZg (Accessed 4 March 2021).

(2020) Dealing with extremism. Online video clip. Available at www.youtube.com/watch?v=aAk0UuzQ31o (Accessed 2 September 2023).

(2023a) *The Handover: How We Gave Control of Our Lives to Corporations, States and AIs*. London: Profile Books.

(2023b) The Handover: How We Gave Control of Our Lives to Corporations, States and AIs – an interview – The Condutit (7 September). Available at www.theconduit.com/past-events/the-handover-how-we-gave-control-of-our-lives-to-corporations-states-and-ais-sep-6th/ (Accessed 15 October 2023).

Russell, Stuart (2019) *Human Compatible: AI and the Problem of Control*. London: Allen Lane.

(2021) Reith Lectures 2021. Available at www.bbc.co.uk/programmes/b00729d9/episodes/player

(2022) Rethinking AI. In A. Rajan (ed.), *Rethink: Leading Voices on Life after Crisis and How to Make a Better World*. London: BBC Books/Penguin, pp. 279–282.

Salam, Erum (2023) US mother gets call from 'kidnapped daughter' – but it's really an AI scam. *The Guardian* (14 June). Available at www.theguardian.com/us-news/2023/jun/14/ai-kidnapping-scam-senate-hearing-jennifer-destefano (Accessed 2 September 2023).

Sample, Ian (2015) Baby girl is first in the world to be treated with 'designer immune cells'. *The Guardian* (5 November). Available at www.theguardian.com/science/2015/nov/05/baby-girl-is-first-in-the-world-to-be-treated-with-designer-immune-cells (Accessed 2 September 2023).

(2023) First UK baby with DNA from three people born after new IVF procedure. *The Guardian* (9 May). Available at www.theguardian.com/science/2023/may/09/first-uk-baby-with-dna-from-three-people-born-after-new-ivf-procedure.

Sandel, Michael (2007). *The Case Against Perfection: Ethics in the Age of Genetic Engineering*. Cambridge, MA: Harvard University Press.

(2020) *The Tyranny of Merit: What's Become of the Common Good?* London: Penguin.

Sandberg, Sheryl (2013) *Lean In. Women, Work and the Will to Lead*. London: Random House Publishing.

(2015) Sandberg and Branson defend Facebook's egg-freezing policy. 24 April interview with Emily Chang. New York: Bloomberg. Available at http://time.com/3835233/sheryl-sandberg-explains-why-facebook-covers-egg-freezing/ (Accessed 4 April 2024).

Sangiovanni, Andrea (2017) *Humanity without Dignity: Moral Equality, Respect, and Human Rights*. Cambridge, MA: Harvard University Press.

(2018) Structural injustice and individual responsibility. *Journal of Social Philosophy* 49(3): 461–483.

Sartre, Jean-Paul (1976) *Critique of Dialectical Reason*, trans. Alan Sheridan-Smith. London: New Left Books.

Saundarya, Rajesh, Kandaswamy, Umasanker and Rakesh, Anju (2018) The impact of artificial intelligence in talent acquisition lifecycle of organizations; a global perspective. *International Journal of Engineering Development and Research* 6(2): 2321–9939.

Savulescu, Julian (2011) The HFEA has restricted liberty without good cause. *The Guardian* (7 February). Available at www.theguardian.com/commentisfree/belief/2011/feb/07/hfea-reproductive-technology-research (Accessed 8 September 2023).

(2014) As a species, we have a moral obligation to enhance ourselves. Interview with Julian Savulescu, Ideas.TED.com. Available at https://ideas.ted.com/the-ethics-of-genetically-enhanced-monkey-slaves/ (Accessed 2 September 2023).

(2020) Genetic enhancement is a moral obligation. The Future of Genetics. Doha Debates (28 May). Available at https://dohadebates.com/video/julian-savulescu-genetic-enhancement-is-a-moral-obligation/ (Accessed 2 September 2023).

Schieffelin, Edward (1980) Reciprocity and the construction of reality. *Man* 15 (3): 502–517.

Schiff, Jacob (2012) The varieties of thoughtlessness and the limits of thinking. *European Journal of Political Theory* 12(2): 99–115.

Schiff, Jade Larissa (2014) *Burdens of Political Responsibility: Narrative and the Cultivation of Responsiveness* (Kindle edition). Cambridge: Cambridge University Press.

Scholl, Inge (1983) *The White Rose: Munich, 1942–1943*. Middletown, CT: Wesleyan University Press.

Schwab, Klaus (2017) *The Fourth Industrial Revolution*. London: Penguin.

Schwenbecher, Anne (2010) How to punish collective agents: Non-compliance with moral duties by states. *Ethics and International Affairs* 24: 1–5.

Sen, Amartya (2004) Elements of a theory of human rights. *Philosophy and Public Affairs* 32(4): 315–356.

Setälä, Maija and Smith, Graham (2018) Mini publics and deliberative democracy. In André Bächtiger, John Dryzek, Jane Mansbridge and Mark Warren (eds.), *The Oxford Handbook of Deliberative Democracy*. Oxford: Oxford University Press, pp. 300–314.

Setälä, Maija, Christensen, Henrik Serup and Leino, Mikko (2021) Beyond polarization and selective trust: A Citizen's Jury as a trusted source of information. *Politics*. Available at https://journals.sagepub.com/doi/pdf/10.1177/02633957211024474 (Accessed 2 September 2023).

Sewell, William H. (2005) *Logics of History: Social Theory and Social Transformations*. Chicago, IL: University of Chicago Press.

Shamoun, Dima Yazji and Yandle, Bruce (2016) Asserting presidential preferences in a regulatory review bureaucracy. *Public Choice* 166(1): 87–111.

Shapiro, I (1999) Enough of deliberation: Politics is about interests and power. In S. Macedo (ed.), *Deliberative Politics: Essays on Democracy and Disagreement*. New York: Oxford University Press, pp. 28–38.

Shapiro, Ian (2016) *Politics Against Domination*. Cambridge, MA: Harvard University Press.

Sherwin, Susan (1987) Feminist ethics and in vitro fertilization. *Canadian Journal of Philosophy* 17(sup1): 264–284.

Shulman, Carl and Bostrom, Nick (2014) Embryo selection for cognitive enhancement: Curiosity or game-changer? *Global Policy* 5(1): 85–92.

(2021) Sharing the world with digital mind. In Steve Clarke, Hazem Zohny and Julian Savulescu (eds.), *Rethinking Moral Status*. Oxford: Oxford University Press, pp. 306–326.

Sikkink, Kathryn (2020) *The Hidden Face of Rights: Toward a Politics of Responsibilities*. New Haven, CT: Yale University Press.

Sillars, James (2023) BT to slash workforce by up to 55,000 before 2030, with AI replacing 10,000 jobs. *Sky News* (18 May). Available at https://news.sky.com/story/bt-aims-to-slash-workforce-by-up-to-55-000-before-2030-12883383 (Accessed 2 September 2023).

Skidelsky, Robert and Skidelsky, Edward (2013) *How Much Is Enough?: Money and the Good Life*. London: Penguin.

Sluga, Hans (2014) *Politics and the Search for the Common Good*. Cambridge: Cambridge University Press.

Smith, Adam (2010 [1759]) *The Theory of Moral Sentiments*. London: Penguin Classics.

(1991 [1776]) *The Wealth of Nations*. London: Penguin.

Smith, Brad (2023) quoted in Zurcher, Anthony (2023) AI: How 'freaked out' should we be?' *BBC News* (16 March). Available at www.bbc.co.uk/news/world-us-canada-64967627 (Accessed 2 September 2023).

Snyder, Timothy (2023) Is democracy doomed? The global fight for our future. Online video clip. Available at www.youtube.com/watch?v=YY6LCOJbve8 (Accessed 2 September 2023).

Stangneth, Bettina (2014) *Eichmann before Jerusalem: The Unexamined Life of a Mass Murderer.* London: Penguin Random House.

Stigler, George, J. (1975) *The Citizen and the State: Essays on Regulation.* Chicago, IL: University of Chicago Press.

Steinberger, Peter J. (1990) Hannah Arendt on judgment. *American Journal of Political Science* 34(3): 803–821.

Strathern, Marilyn (1992) *The Gender of the Gift: Problems with Women and Problems with Society.* Berkeley: University of California Press.

Suleyman, Mustafa (2023) *The Coming Wave.* London: Penguin Random House.

Sunstein, Cass and Vermeule, Adrian (2020) *Law and Leviathan: Redeeming the Administrative State.* Cambridge, MA: Belknap Press.

Sutton Trust (2022) Nearly two-thirds of new cabinet attended independent schools and almost half attended Oxbridge. Available at www.suttontrust.com/news-opinion/all-news-opinion/nearly-two-thirds-of-new-cabinet-attended-independent-schools-and-almost-half-attended-oxbridge/ (Accessed 2 September 2023).

Taddeo, M. and Floridi L. (2018) How AU can be a force for good. *Science* 361 (6404): 751–752.

Tarasoff, Lesley, Cattapan, Alana, Hammond, Kathleen and Haw, Jennie (2014) Breaking the ice: Young feminist scholars of reproductive politics reflect on egg freezing. *International Journal of Feminist Approaches to Bioethics* 7: 236–247.

Taylor, Astra (2019) *Democracy May Not Exist but We'll Miss It When It's Gone.* New York: Metropolitan Books.

Taylor, Charles (1989) Cross purposes: The liberal-communitarian debate. In N. Rosenblum (ed.), *Liberalism and the Moral Life.* Cambridge, MA: Harvard University Press, pp. 159–182.

The Economist (2017) Adidas's high-tech factory brings production back to Germany (14 January). Available at www.economist.com/business/2017/01/14/adidass-high-tech-factory-brings-production-back-to-germany (Accessed 2 September 2023).

Thunberg, Greta (2018) Greta Thunberg addresses the UN Climate Change COP24 Conference in Poland. Connect4Climate. Online video clip. Available at www.youtube.com/watch?v=VFkQSGyeCWg (Accessed 2 September 2023).

(2021) quoted in BBC (2021): Who is the climate campaigner and what are her aims? (5 November). Available at www.bbc.co.uk/news/world-europe-49918719 (Accessed 2 September 2023).

Time Magazine Cover (2011) 2045 – The year man becomes immortal. Available at http://content.time.com/time/covers/0,16641,20110221,00.html (Accessed 2 September 2023).

An Tionól Saoránach/Citizens' Assembly (2017) First Report and Recommendations of the Citizens' Assembly; The Eighth Amendment of the Constitution (29 June). Available at https://citizensassembly.ie/wp-con tent/uploads/2023/02/FirstReport_EIGHTAMENDMENT.pdf (Accessed 2 September 2023).

Torres, Emile P. (2023) The acronym behind our wildest AI dreams and night-mares. Truthdig (15 June). Available at www.truthdig.com/dig-series/eugen ics/ (Accessed 2 September 2023).

Trades Union Congress (TUC) (2020) Insecure work: Why decent work needs to be at the heart of the UK's recovery from coronavirus (8 August). Available at www.tuc.org.uk/research-analysis/reports/insecure-work-0 (Accessed 2 September 2023).

Trnka, Daniel and Thuerer, Yola (2019) One-In, X-Out: Regulatory offsetting in selected OECD countries. OECD Regulatory Policy Working Papers, No. 11, Paris: OECD Publishing. https://doi.org/10.1787/67d71764-en.

Trump, Donald (2016) First 100 days plan. CNN News, USA (21 November). Available at http://edition. cnn.com/2016/11/21/politics/donald-trump-outlines-policy-plan-for-first-100-days/ (Accessed 2 September 2023).

Turing, A. (1950) Computing machinery and intelligence. *Mind* 49: 433–460.

Turner, Bryan S. (2019) Max Weber and the tragedy of politics: Reflections on unintended consequences of action. *Journal of Classical Sociology* 19(4): 377–390.

UK Government – Maternity Pay & Leave (2024). Available at www.gov.uk/mater nity-pay-leave/pay#:~:text=Statutory%20Maternity%20Pay%20(%20SMP%20) %20is,for%20the%20next%2033%20weeks (Accessed 1 April 2024).

UK Research and Innovation (UKRI) (2022) A brief history of climate change discoveries. Available at www.discover.ukri.org/a-brief-history-of-climate-change-discoveries/index.html (Accessed 2 September 2023).

UK Research and Innovation (UKRI) (2024) UKRI artificial intelligence Centres for Doctoral Training. Available at www.ukri.org/who-we-are/our-vision-and-strategy/tomorrows-technologies/how-we-work-in-ai/ukri-artificial-intelli gence-centres-for-doctoral-training/#:~:text=UKRI%20is%20investing%20a% 20further,proposals%20by%2013%20July%202023 (Accessed 4 April 2024).

UNESCO (2021) General Conference, 41st. Report of the Social and Human Sciences Commission (SHS). Available at https://unesdoc.unesco.org/ark:/48223/pf0000379920.page=14 (Accessed 2 September 2023).

United Nations (UN) (2021a) World population prospects. Available at https://population.un.org/wpp/ (Accessed 2 September 2023).

(2021b) The Convention on Certain Conventional Weapons. Available at www.un.org/disarmament/the-convention-on-certain-conventional-weapons/ (Accessed 2 September 2023).

US Bureau of Labour Statistics (2023) News Release: Employment Situation – August 2023. Available at www.bls.gov/news.release/pdf/empsit.pdf (Accessed 3 September 2023).

US District Court for the District of Columbia (2018) Public Citizen, Inc., Natural Resources Defense Council, Inc., and Communications Workers of America, AFL-CIO, United States District Court for the District of Columbia. Case 1:17-cv-00253-RDM Document 67 Filed 04/20/18. Second Amended Complaint for Declaratory and Injunctive Relief. Available at https://mkus3lurbh3lbztg254fzode-wpengine.netdna-ssl.com/wp-content/uploads/second-amended-complaint-public-citizen-trump.pdf (Accessed 2 September 2023).

US Federal Government (2023) National Artificial Intelligence Research and Development Strategic Plan 2023 Update. Available at chrome-extension://efaidnbmnnnibpcajpcglclefindmkaj/https://www.whitehouse.gov/wp-content/uploads/2023/05/National-Artificial-Intelligence-Research-and-Development-Strategic-Plan-2023-Update.pdf (Accessed 9 April 2024).

van de Wiel, Lucy (2020a) The speculative turn in IVF: Egg freezing and the financialization of fertility. *New Genetics and Society* 39(3): 306–326.

(2020b) Freezing Fertility: Oocyte Cryopreservation and the Gender Politics of Aging [Internet]. New York: New York University Press. Available at www.ncbi.nlm.nih.gov/books/NBK568241/ (Accessed 6 September 2023).

van Leeuwen, Flora, Klip, Helen, Mooij, Thea, van de Swaluw, Jojanneke, Lambalk, Cornelis, Kortman, Marianne, Laven, Joop, et al. (2011). Risk of borderline and invasive ovarian tumours after ovarian stimulation for in vitro fertilization in a large Dutch cohort. *Human Reproduction* 26(12): 3456–65.

Villani, Cédric (2018) For a Meaningful Artificial Intelligence: Towards a French and European Strategy. Mission assigned by the Prime Minister Édouard Philippe. A parliamentary mission from 8th September 2017 to 8th March 2018. Available at www.aiforhumanity.fr/pdfs/MissionVillani_Report_ENG-VF.pdf (Accessed 2 September 2023).

Von Eschenbach, W. J. (2021) Transparency and the black box problem: Why we do not trust AI. *Philosophy & Technology* 34: 1607–1622.

Wagner, Wendy (2010) Administrative law, filter failure, and information capture. *Duke Law Journal* 59, 1321–1432.

Waldby, Catherine (2015) The oocyte market and social egg freezing: From scarcity to singularity. *Journal of Cultural Economy* 8(3):275–291.

Walzer, Michael (1999) Deliberation, and what else? In Stephen Macedo (ed.), *Deliberative Politics: Essays on Democracy and Disagreement.* New York: Oxford University Press, pp. 58–69.

Wang, Xiaoyu (2021) Population gets older as growth slows. The State Council: The People's Republic of China. Available at http://english.www.gov.cn/statecouncil/ministries/202105/12/content_WS609b1523c6d0df57f98d9600.html (Accessed 2 September 2023).

Warnock, Mary (1984) Report of the Committee of Inquiry into Human Fertilisation and Embryology. London: HMSO, Wellcome Collection. Available at https://wellcomecollection.org/works/pxgeeqnf (Accessed 2 September 2023).

Watson, David and Floridi, Luciano (2020) The explanation game: a formal framework for interpretable machine learning. Synthese. Available at https://link.springer.com/article/10.1007/s11229–020–02629-9#citeas (Accessed 2 September 2023).

Weatherbee, B. A. T., Gantner, C. W., Iwamoto-Stohl L. K. et al. (2023) Pluripotent stem cell-derived model of the post-implantation human embryo. *Nature.* https://doi.org/10.1038/s41586–023-06368-y.

Weber, Max (2002 [1919]) *The Profession and Vocation of Politics in Max Weber,* ed. Peter Lassman and Ronal Speirs. Political Writings. Cambridge: Cambridge University Press, pp. 309–369.

    (1919) Politics as a vocation. In H. H. Gerth and C. Wright Mills (trans. and ed.) (1946), *Essays in Sociology.* New York: Oxford University Press, pp. 77–128.

Webb, Amy (2023) launches 2023 emerging tech trend report. Online video clip. Available at SXSW www.youtube.com/watch?v=vMUpzxZB3-Y (Accessed 2 September 2023).

Williams, Chris (2015) AI guru Ng: Fearing a rise of killer robots is like worrying about overpopulation on Mars. *The Register* (19 March). Available at www.theregister.com/2015/03/19/andrew_ng_baidu_ai/#:~:text=GTC%202015%20Artificial%20intelligence%20boffin,even%20set%20foot%20on%20it (Accessed 2 September 2023).

Williams, Matthew (2020) Global solidarity, global worker empowerment, and global strategy in the antisweatshop movement. *Labor Studies Journal* 45(4): 394–420.

Williamson, Gavin (May 2018). Speech at the US-UK Defence Innovation Board Meeting to launch new flagship artificial intelligence lab. Available at www

.gov.uk/government/news/flagship-ai-lab-announced-as-defence-secretary-hosts-first-meet-between-british-and-american-defence-innovators (Accessed 2 September 2023).

Winchester, Simon (2023) *Knowing What We Know: The Transmission of Knowledge: From Ancient Wisdom to Modern Magic.* Glasgow: William Collins Publisher.

Wired (2022) 'I'm the Operator': The Aftermath of a Self-Driving Tragedy: Lauren Smiley (8 March). Available at www.wired.com/story/uber-self-driving-car-fatal-crash/ (Accessed 2 September 2023).

Wolff, Jonathan (2018) Structural injustice. Centre for the Study of Global Ethics, public lecture at the University of Birmingham (31 May). Online video clip. Available at www.youtube.com/watch?v=dppTGpcVGHc (Accessed 2 September 2023).

  (2024) Structural harm, structural injustice, structural repair. In Jude Browne and Maeve McKeown (eds.), *What Is Structural Injustice?* Oxford: Oxford University Press, pp. 12–30.

World Economic Forum (2018) The Future of Jobs 2018 Report. Available at http://reports.weforum.org/future-of-jobs-2018/ (Accessed 2 September 2023).

  (2020) The Future of Jobs 2020 Report. Available at https://www3.weforum .org/docs/WEF_Future_of_Jobs_2020.pdf (Accessed 2 September 2023).

  (2022) Without universal AI literacy, AI will fail us (17 March). Available at www.weforum.org/agenda/2022/03/without-universal-ai-literacy-ai-will-fail-us/ (Accessed 2 September 2023).

World Population Review (2023) Total fertility rate. Available at https:// worldpopulationreview.com/country-rankings/total-fertility-rate (Accessed 2 September 2023).

Young, Iris Marion (1990) *Justice and the Politics of Difference.* Princeton, NJ: Princeton University Press.

  (1996) Communication and the other: Beyond deliberative democracy. In Seyla Benhabib (ed.), *Democracy and Difference: Contesting the Boundaries of the Political.* Princeton, NJ: Princeton University Press, pp. 120–136.

  (1997) *Intersecting Voices: Dilemmas of Gender, Political Philosophy and Policy.* Princeton, NJ: Princeton University Press.

  (2000) *Inclusion and Democracy.* New York and Oxford: Oxford University Press.

  (2003) The Lindley Lecture. University of Kansas (5 May). Available at https:// kuscholarworks.ku.edu/bitstream/handle/1808/12416/politicalresponsibili tyandstructuralinjustice-2003.pdf?sequence=1 (Accessed 2 September 2023).

  (2006) Responsibility and global justice: A social connection model. *Social Philosophy and Policy* 23(1): 102–130.

(2009) Structural injustice and the politics of difference. In Thomas Christiano and John Christman (eds.), *Contemporary Debates in Political Philosophy: 11*. Chichester: Wiley-Blackwell, pp. 362–384.

(2011) *Responsibility for Justice*. Oxford: Oxford University Press.

Yosinski, Jason (2017) How AI detectives are cracking open the black box of deep learning. *Science*. Available at http://science.sciencemag.org/ (Accessed 2 September 2023).

Yudkowsky, Eliezer (2023) Pausing AI developments is not enough. We need to shut it all down. *Time* (29 March). Available at https://time.com/6266923/ai-eliezer-yudkowsky-open-letter-not-enough/ (Accessed 2 September 2023).

Zheng, Robin (2018) What is my role in changing the system? A new model of responsibility for structural injustice. *Ethical Theory and Moral Practice* 21: 869–885.

(2019) What kind of responsibility do we have for fighting injustice? A moral-theoretic perspective on the social connections model. *Critical Horizons* 20 (2): 109–126.

(2021) Moral criticism and structural injustice. *Mind* 130(518): 503–535.

Zihao, Li, Yang, Zhuoran and Mengdi, Wang (2023) Reinforcement Learning with Human Feedback: Learning Dynamic Choices via Pessimism. Cornell University. Available at https://arxiv.org/pdf/2305.18438.pdf (Accessed 2 September 2023).

Zingales, Luigi (2014) Preventing economists' capture. In David Capenter and David, A. Moss (eds.), *Preventing Regulatory Capture: Special Interest Influence and How to Limit It*. Cambridge: Cambridge University Press. pp. 124–151.

Zinn, Howard (1995) *A People's History of the United States*. New York: HarperCollins.

Zupan, Mark (2017) *Inside Job: How Government Insiders Subvert the Public Interest*. Cambridge: Cambridge University Press.

Index

For EU product safety concerns, contact us at Calle de José Abascal, 56–1°, 28003 Madrid, Spain or eugpsr@cambridge.org.